Selected Titles in This Series

89 **Michel Ledoux,** The concentration of measure phenomenon, 2001
88 **Edward Frenkel and David Ben-Zvi,** Vertex algebras and algebraic curves, 2001
87 **Bruno Poizat,** Stable groups, 2001
86 **Stanley N. Burris,** Number theoretic density and logical limit laws, 2001
85 **V. A. Kozlov, V. G. Maz′ya, and J. Rossmann,** Spectral problems associated with corner singularities of solutions to elliptic equations, 2001
84 **László Fuchs and Luigi Salce,** Modules over non-Noetherian domains, 2001
83 **Sigurdur Helgason,** Groups and geometric analysis: Integral geometry, invariant differential operators, and spherical functions, 2000
82 **Goro Shimura,** Arithmeticity in the theory of automorphic forms, 2000
81 **Michael E. Taylor,** Tools for PDE: Pseudodifferential operators, paradifferential operators, and layer potentials, 2000
80 **Lindsay N. Childs,** Taming wild extensions: Hopf algebras and local Galois module theory, 2000
79 **Joseph A. Cima and William T. Ross,** The backward shift on the Hardy space, 2000
78 **Boris A. Kupershmidt,** KP or mKP: Noncommutative mathematics of Lagrangian, Hamiltonian, and integrable systems, 2000
77 **Fumio Hiai and Dénes Petz,** The semicircle law, free random variables and entropy, 2000
76 **Frederick P. Gardiner and Nikola Lakic,** Quasiconformal Teichmüller theory, 2000
75 **Greg Hjorth,** Classification and orbit equivalence relations, 2000
74 **Daniel W. Stroock,** An introduction to the analysis of paths on a Riemannian manifold, 2000
73 **John Locker,** Spectral theory of non-self-adjoint two-point differential operators, 2000
72 **Gerald Teschl,** Jacobi operators and completely integrable nonlinear lattices, 1999
71 **Lajos Pukánszky,** Characters of connected Lie groups, 1999
70 **Carmen Chicone and Yuri Latushkin,** Evolution semigroups in dynamical systems and differential equations, 1999
69 **C. T. C. Wall (A. A. Ranicki, Editor),** Surgery on compact manifolds, second edition, 1999
68 **David A. Cox and Sheldon Katz,** Mirror symmetry and algebraic geometry, 1999
67 **A. Borel and N. Wallach,** Continuous cohomology, discrete subgroups, and representations of reductive groups, second edition, 2000
66 **Yu. Ilyashenko and Weigu Li,** Nonlocal bifurcations, 1999
65 **Carl Faith,** Rings and things and a fine array of twentieth century associative algebra, 1999
64 **Rene A. Carmona and Boris Rozovskii, Editors,** Stochastic partial differential equations: Six perspectives, 1999
63 **Mark Hovey,** Model categories, 1999
62 **Vladimir I. Bogachev,** Gaussian measures, 1998
61 **W. Norrie Everitt and Lawrence Markus,** Boundary value problems and symplectic algebra for ordinary differential and quasi-differential operators, 1999
60 **Iain Raeburn and Dana P. Williams,** Morita equivalence and continuous-trace C^*-algebras, 1998
59 **Paul Howard and Jean E. Rubin,** Consequences of the axiom of choice, 1998

For a complete list of titles in this series, visit the AMS Bookstore at **www.ams.org/bookstore/**.

Mathematical
Surveys
and
Monographs

Volume 89

The Concentration of Measure Phenomenon

Michel Ledoux

American Mathematical Society
Providence, Rhode Island

2000 *Mathematics Subject Classification.* Primary 28Axx, 46Bxx, 52Axx, 60–XX; Secondary 28C20, 28D20, 46G12, 58C30, 62G30, 82B44.

ABSTRACT. This book presents the basic aspects of the concentration of measure phenomenon that was put forward in the early seventies, and emphasized since then, by V. Milman in asymptotic geometric analysis. It has now become of powerful interest in applications, in various areas such as geometry, functional analysis and infinite dimensional integration, discrete mathematics and complexity theory, and probability theory. The book is concerned with the basic techniques and examples of the concentration of measure phenomenon. A particular emphasis has been put on geometric, functional and probabilistic tools to reach and describe measure concentration in a number of settings, as well as on M. Talagrand's investigation of concentration in product spaces and its application in discrete mathematics and probability theory.

Library of Congress Cataloging-in-Publication Data

Ledoux, Michel, 958–
The concentration of measure phenomenon / Michel Ledoux.
p. cm. — (Mathematical surveys and monographs, ISSN 0076-5376 ; v. 89)
Includes bibliographical references and index.
ISBN 0-8218-2864-9 (acid-free paper)
1. Concentration functions. 2. Measure theory. 3. Probabilities. I. Title. II. Mathematical surveys and monographs ; no. 89.

QA273.43.L43 2001
519.2—dc21 2001041310

AMS soft cover ISBN: 978-0-8218-3792-4

∞ The paper used in this book is acid-free and falls within the guidelines
established to ensure permanence and durability.
Visit the AMS home page at URL: http://www.ams.org/

13 12 11 10 9 8 7 6 5 26 25 24 23 22 21

CONTENTS

INTRODUCTION

The aim of this book is to present the basic aspects of the concentration of measure phenomenon. The concentration of measure phenomenon was put forward in the early seventies by V. Milman in the asymptotic geometry of Banach spaces. Of isoperimetric inspiration, it is of powerful interest in applications, in various areas such as geometry, functional analysis and infinite dimensional integration, discrete mathematics and complexity theory, and especially probability theory. This book is concerned with the basic techniques and examples of the concentration of measure phenomenon with no claim to be exhaustive. A particular emphasis has been put on geometric, functional and probabilistic tools to reach and describe measure concentration in a number of settings.

As mentioned by M. Gromov, the concentration of measure phenomenon is an elementary, yet non-trivial, observation. It is often a high dimensional effect, or a property of a large number of variables, for which functions with small local oscillations are almost constant. A first illustration of this property is suggested by the example of the standard n-sphere $\mathbb{S}^n$ in $\mathbb{R}^{n+1}$ when the dimension n is large. One striking aspect of uniform measure σ^n on $\mathbb{S}^n$ in high dimension is that it is almost concentrated around the equator. More generally, as a consequence of spherical isoperimetry, given any measurable set A with, say, $\sigma^n(A) \geq \frac{1}{2}$, almost all points (in the sense of the measure σ^n) on $\mathbb{S}^n$ are within (geodesic) distance $\frac{1}{\sqrt{n}}$ from A (which becomes infinitesimal as $n \to \infty$). Precisely, for every $r > 0$,

$$\sigma^n(A_r) \geq 1 - \mathrm{e}^{-(n-1)r^2/2}$$

where $A_r = \{x \in \mathbb{S}^n; d(x, A) < r\}$ is the neighborhood of order $r > 0$ of A for the geodesic metric on $\mathbb{S}^n$.

This concentration property on the sphere may be described equivalently on functions, an idea going back to Lévy. Namely, if F is a continuous function on $\mathbb{S}^n$ with modulus of continuity $\omega_F(\eta) = \sup\{|F(x) - F(y)|; d(x, y) < \eta\}$, then

$$\sigma^n\big(\{|F - m_F| \geq \omega_F(\eta)\}\big) \leq 2\,\mathrm{e}^{-(n-1)\eta^2}$$

where m_F is a median of F for σ^n. Therefore, functions on high dimensional spheres with small local oscillations are strongly concentrated around a mean value, and are thus almost constant on almost all the space! This high dimensional concentration phenomenon was extensively used and emphasized by V. Milman in his investigation of asymptotic geometric analysis.

As yet another interpretation, the concept of observable diameter as considered by M. Gromov is a "visual" description of the concentration of measure phenomenon. We view the sphere with a naked eye which cannot distinguish a part of $\mathbb{S}^n$ of measure (luminosity) less than $\kappa > 0$ (small but fixed). A Lipschitz function F may be interpreted as an observable, that is an observation device giving us the visual image measure of σ^n by F. In this language, Lévy's inequality on Lipschitz functions expresses that the "observable diameter of $\mathbb{S}^n$" is of the order of $\frac{1}{\sqrt{n}}$ as n is large, in strong contrast with the diameter of $\mathbb{S}^n$ as a metric space.

In probability theory, the concentration of measure is a property of a large number of varibales, such as in laws of large numbers. A probabilistic description of the concentration phenomenon goes back to E. Borel who suggested the following geometric interpretation of the law of large numbers for sums of independent random variables uniformly distributed on the interval $[0, 1]$. Let μ^n be uniform measure on the n-dimensional cube $[0, 1]^n$. Let H be a hyperplane that is orthogonal to a principal diagonal of $[0, 1]^n$ at the center of the cube. Then, if H_r is the neighborhood of order $r > 0$ of H, for every $\varepsilon > 0$, $\mu^n(H_{\varepsilon\sqrt{n}}) \to 1$ as $n \to \infty$. Actually, the relevant observable is $x \in [0, 1]^n \mapsto \frac{1}{n}\sum_{i=1}^n x_i \in [0, 1]$, which concentrates around the mean value $\frac{1}{2}$. The normal projection of the cube to the principal diagonal identified with $[0, \sqrt{n}]$ thus sends most of the measure of the cube to the subsegment

$$\left[\frac{\sqrt{n}}{2} - \varepsilon\sqrt{n}, \frac{\sqrt{n}}{2} + \varepsilon\sqrt{n}\right].$$

In fact, $\varepsilon\sqrt{n}$ may be replaced by any sequence $r_n \to \infty$ as follows from the central limit theorem.

A related description is the following. Let $X_1, X_2, \dots$ be a sequence of independent random variables taking the values ± 1 with equal probability, and set, for every $n \geq 1$, $S_n = X_1 + \cdots + X_n$. We think of S_n as a function of the individual variables X_i and we state the classical law of large numbers by saying that S_n is essentially constant (equal to 0). Of course, by the central limit theorem, the fluctuations of S_n are of order $\sqrt{n}$ which is hardly zero. But as S_n can take values as large as n, this is the scale at which one should measure S_n, in which case S_n/n is indeed essentially zero as expressed by the classical exponential bound

$$\mathbb{P}\bigg(\bigg\{\frac{|S_n|}{n} \geq r\bigg\}\bigg) \leq 2\,\mathrm{e}^{-nr^2/2}, \quad r \geq 0.$$

In this context, and according to M. Talagrand, one probabilistic aspect of measure concentration is that a random variable that depends (in a smooth way) on the influence of many independent variables (but not too much on any of them) is essentially constant.

Measure concentration is surprisingly shared by a number of cases that generalize the previous examples, both by replacing linear functionals (such as sums of independent random variables) by arbitrary Lipschitz functions of the samples, and by considering measures that are not of product form. It was indeed again the insight of V. Milman to emphasize the difference between the concentration phenomenon and standard probabilistic views on probability inequalities and law of large number theorems by the extension to Lipschitz (and even Hölder type) functions and more general measures. His enthusiasm and persuasion eventually convinced M. Talagrand of the importance of this simple, yet fundamental, concept.

It will be one of the purposes of this book to describe some of the basic examples and applications of the concentration of measure phenomenon. While the first applications were mainly developed in the context of asymptotic geometric analysis, they have now spread to a wide range of frameworks, covering areas in geometry, discrete and combinatorial mathematics, and in particular probability theory. Classical probabilistic inequalities on sums of independent random variables have been used indeed over the years in limit theorems and discrete algorithmic mathematics. They provide quantitative illustrations of measure concentration by so-called exponential inequalities (mostly of Gaussian type). Recent developments on the concentration of measure phenomenon describe far reaching extensions that provide dimension free concentration properties in product spaces which, due to the work of M. Talagrand during the last decade, will form a main part of these notes.

The book is divided into 8 chapters. The first one introduces the notions and elementary properties of concentration functions, deviation inequalities and their more geometric counterparts as observable diameters. We also briefly indicate a few useful tools to investigate concentration properties. The second chapter describes some of the basic and classical isoperimetric inequalities at the origin of the concentration of measure phenomenon. However, we do not concentrate on the usually somewhat delicate extremal statements, but rather develop some self-contained convexity and semigroup arguments to reach the concentration properties originally deduced from isoperimetry. Chapter 3 is a first view towards geometric and topological applications of measure concentration. In particular, we describe there Milman's proof of Dvoretsky's theorem on almost spherical sections of convex bodies. V. Milman in this proof most vigorously emphasized the usefulness of concentration ideas. Chapter 4 investigates measure concentration in product spaces, mostly based on the recent developments by M. Talagrand. After a brief view of the more classical martingale bounded difference method, we cover there the convex hull and finite point approximations, which are of powerful use in applications to both empirical processes and discrete mathematics. We also discuss the particular concentration property of the exponential distribution. The next two chapters emphasize functional inequalities stable under products thereby obtaining a new approach to the results of Chapter 4. Chapter 5 is devoted to the entropic and logarithmic Sobolev inequality approach. We present there the Herbst method to deduce concentration from a logarithmic Sobolev inequality and describe the various applications to product measures and related topics. Chapter 6 is yet another form of concentration relying on information and transportation cost inequalities with which one may reach several of the conclusions of the preceding chapters. Chapter 7 is devoted to the probabilistic applications of concentration in product spaces to sharp bounds on sums of independent random vectors or empirical processes: these applications lay at the heart of M. Talagrand's original investigation. The last chapter is a selection of (recent) applications of the concentration of measure phenomenon to various areas such as statistical mechanics, geometric probabilities, discrete and algorithmic mathematics, for which the concentration ideas, although perhaps at some mild level, appear to be useful tools of investigation.

While we describe in this work a number of concentration properties put forward in several contexts, from more geometric to functional and probabilistic settings, we usually produce the correct orders but almost never discuss sharp constants.

This book is strongly inspired by early references on the subject. In particular, the lecture notes by V. Milman and G. Schechtman that describe the concentration of measure phenomenon and its applications to asymptotic theory of finite dimensional normed spaces were a basic source of inspiration during the preparation of this book. We also used the recent survey by G. Schechtman in the Handbook in the Geometry of Banach Spaces. (The latter handbook contains further contributions that illustrate the use of concentration in various functional analytic problems.) The memoir by M. Talagrand on isoperimetric and concentration inequalities in product spaces is at the basis of most of the material presented starting with Chapter 4, and the ideas developed there gave a strong impetus to recent developments in various areas of probability theory and its applications. Several of the neat arguments presented in these references have been reproduced here. The already famous $3\frac{1}{2}$ Chapter of the recent book by M. Gromov served as a useful source of geometric examples where further motivating aspects of convergence of metric measure spaces related to concentration are developed. While many geometric invariants are introduced and analyzed there, our point of view is perhaps a bit more quantitative and motivated by a number of recent probabilistic questions. Perspectives and developments related to the concentration of measure phenomenon in various areas of mathematics and its applications are discussed in M. Gromov's book as well as in the recent papers of M. Gromov and V. Milman in the special issues "Vision" of Geometric and Functional Analysis GAFA2000.

Each chapter is followed by some Notes and Remarks with an attempt in particular to trace the origin of the main ideas. We apologize for inaccuracies and omissions.

The notations used throughout this book are the standard ones used in the literature. Although we keep some consistency, we did not try to unify all the notations and often used the classical notation in a given context even though it might have been used differently in another.

I am grateful to Michel Talagrand for numerous discussions over the years on the topic of concentration and for explaining to me his work on concentration in product spaces. Parts of several joint works with Sergey Bobkov on concentration and related matters are reproduced here. I sincerely thank him for corrections and comments on the first draft of the manuscript. I also thank Vitali Milman and Gideon Schechtman for their interest and useful comments and suggestions, and Markus Neuhauser, James Norris and Vladimir Pestov for helpful remarks and corrections. I sincerely thank the A.M.S. Mathematics Editor Edward Dunne and Natalya Pluzhnikov for their help in the preparation of the manuscript.

Toulouse, May 2001 Michel Ledoux

1. CONCENTRATION FUNCTIONS AND INEQUALITIES

In this chapter, we introduce, with the first examples of spherical and Gaussian isoperimetry, the concept of measure concentration as put forward by V. Milman [M-S] and discuss its first properties. We define the notion of concentration function and connect it with Lévy's deviation and concentration inequalities for Lipschitz functions that provide a main tool in applications. The notion of observable diameter is another more geometric view to concentration. The last two sections of this chapter are devoted to the useful tools of expansion coefficients, Laplace bounds and infimum-convolution inequalities to explore concentration properties.

1.1 First examples

To introduce to the concept of measure concentration, we first briefly discuss a few examples that will be further analyzed (with references) later on.

Our first illustration is suggested by the example of the standard n-sphere $\mathbb{S}^n$ in $\mathbb{R}^{n+1}$ when dimension n is large. By a standard computation, uniform measure σ^n on $\mathbb{S}^n$ is almost concentrated when the dimension n is large around the (every!) equator. Actually, the isoperimetric inequality on $\mathbb{S}^n$ expresses that spherical caps (geodesic balls) minimize the boundary measure at fixed volume. In its integrated form (see Section 2.1), given a Borel set A on $\mathbb{S}^n$ with the same measure as a spherical cap B, then for every $r > 0$,

$$\sigma^n(A_r) \geq \sigma^n(B_r)$$

where $A_r = \{x \in \mathbb{S}^n; d(x, A) < r\}$ is the (open) neighborhood of order r for the geodesic distance on $\mathbb{S}^n$. One main feature of concentration with respect to isoperimetry is to analyze this inequality for the non-infinitesimal values of $r > 0$. The explicit evaluation of the measure of spherical caps (performed below in Section 2.1) then implies that given any measurable set A with, say, $\sigma^n(A) \geq \frac{1}{2}$, for every $r > 0$,

$$\sigma^n(A_r) \geq 1 - \mathrm{e}^{-(n-1)r^2/2}. \tag{1.1}$$

Therefore, almost all points on $\mathbb{S}^n$ are within (geodesic) distance $\frac{1}{\sqrt{n}}$ from A, which is of particular interest when the dimension n is large. From a "tomographic" point of view (developed further in Section 1.4 below), the visual diameter of $\mathbb{S}^n$ (for σ^n) is of the order of $\frac{1}{\sqrt{n}}$ as $n \to \infty$, which is in contrast with the diameter of $\mathbb{S}^n$ as a metric space.

This example is a first, and main, instance of the concentration of measure phenomenon for which nice patterns develop as the dimension is large. It furthermore suggests the introduction of a concentration function in order to evaluate the decay in (1.1). Setting

$$\alpha_{\sigma^n}(r) = \sup\big\{1 - \sigma^n(A_r); A \subset \mathbb{S}^n, \sigma^n(A) \geq \tfrac{1}{2}\big\}, \quad r > 0,$$

the bound (1.1) amounts to saying that

$$\alpha_{\sigma^n}(r) \leq \mathrm{e}^{-(n-1)r^2/2}, \quad r > 0. \tag{1.2}$$

Note that $r > 0$ in (1.2) actually ranges up to the diameter π of $\mathbb{S}^n$ and that (1.2) is thus mainly of interest when n is large.

By rescaling of the metric, the preceding results apply similarly to uniform measure σ_R^n on the n-sphere $\mathbb{S}_R^n$ of radius $R > 0$. In particular,

$$\alpha_{\sigma_R^n}(r) \leq \mathrm{e}^{-(n-1)r^2/2R^2}, \quad r > 0. \tag{1.3}$$

Properly normalized, uniform measures on high dimensional spheres approximate Gaussian distributions. More precisely (see Section 2.1), the measures $\sigma_{\sqrt{n}}^n$ converge when n tends to infinity to the canonical Gaussian measure on $\mathbb{R}^{\mathbb{N}}$. The isoperimetric inequality on spheres may then be transferred to an isoperimetric inequality for Gaussian measures. Precisely, if $\gamma = \gamma^k$ is the canonical Gaussian measure on $\mathbb{R}^k$ with density $(2\pi)^{-k/2}\mathrm{e}^{-|x|^2/2}$ with respect to Lebesgue measure and if A is a Borel set in $\mathbb{R}^k$ with $\gamma(A) = \Phi(a)$ for some $a \in [-\infty, +\infty]$ where $\Phi(t) = (2\pi)^{-1/2}\int_{-\infty}^t \mathrm{e}^{-x^2/2}dx$ is the distribution function of the standard normal distribution on the line, then for every $r > 0$,

$$\gamma(A_r) \geq \Phi(a + r).$$

Here A_r denotes the r-neighborhood of A with respect to the standard Euclidean metric on $\mathbb{R}^k$. Unless otherwise specified, $\mathbb{R}^k$ (or subsets of $\mathbb{R}^k$) will be equipped throughout this book with the standard Euclidean structure and the metric $|x - y|$, $x, y \in \mathbb{R}^k$, induced by the norm $|x| = \big(\sum_{i=1}^k x_i^2\big)^{1/2}$, $x = (x_1, \ldots, x_k) \in \mathbb{R}^k$. The scalar product will be denoted $x \cdot y = \sum_{i=1}^k x_i y_i$, $x = (x_1, \ldots, x_k)$, $y = (y_1, \ldots, y_k)$ $\in \mathbb{R}^k$. Defining similarly the concentration function for γ as

$$\alpha_\gamma(r) = \sup\big\{1 - \gamma(A_r); A \subset \mathbb{R}^k, \gamma(A) \geq \tfrac{1}{2}\big\}$$

we get in particular since $\Phi(0) = \frac{1}{2}$ and $1 - \Phi(r) \leq \mathrm{e}^{-r^2/2}$, $r > 0$, that

$$\alpha_\gamma(r) \leq \mathrm{e}^{-r^2/2}, \quad r > 0. \tag{1.4}$$

One may also think of (1.3) in the limit as $n \to \infty$ with $R = \sqrt{n}$. One may again interpret (1.4) by saying that given a set A with $\gamma(A) \geq \frac{1}{2}$, almost all points in $\mathbb{R}^k$ are within distance 5 or 10 say from the set A whereas of course $\mathbb{R}^k$ is unbounded. We have thus here a second instance of measure concentration with the particular feature that the concentration function of (1.4) does not depend on the dimension of the underlying state space $\mathbb{R}^k$ for the product measure $\gamma = \gamma^k$.

Our third example will be discrete. Consider the n-dimensional discrete cube $X = \{0,1\}^n$ and equip X with the normalized Hamming metric

$$d(x,y) = \frac{1}{n}\,\mathrm{Card}\big(\{x_i \neq y_i; i = 1,\dots,n\}\big),$$

$x = (x_1,\dots,x_n), y = (y_1,\dots,y_n) \in \{0,1\}^n$. Let $\mu = \mu^n$ be uniform (product) measure on $\{0,1\}^n$ defined by $\mu(A) = 2^{-n}\,\mathrm{Card}\,(A)$ for every subset A of X. Identifying the extremal sets A in X for which the infimum $\inf\{\mu(A_r); \mu(A) \geq \frac{1}{2}\}$ is attained may be used to show here that

$$\alpha_\mu(r) \leq \mathrm{e}^{-2nr^2}, \quad r > 0, \tag{1.5}$$

where the concentration function α_μ for μ on $\{0,1\}^n$ equipped with the Hamming metric is defined as above.

These first examples of concentration properties all follow from more refined isoperimetric inequalities. They will be detailed in the next chapter in which full proofs of the concentration (rather than isoperimetric) results will be presented. These examples will serve as guidelines for the further developments. They motivate and justify in particular the analysis of the concept of concentration function performed in the next section.

1.2 Concentration functions

Motivated by the early examples of the preceding section, we introduce and formalize the concept of concentration function of a probability measure on, say, a metric space. The concentration examples of Section 1.1 indeed rely on two main ingredients, a (probability) measure and a notion of (isoperimetric) enlargement with respect to which concentration is evaluated.

Thus, let (X,d) be a metric space equipped with a probability measure μ on the Borel sets of (X,d) (a *metric measure space* in the sense of [Grom2]). The *concentration function* $\alpha_{(X,d,\mu)}$ (denoted more simply $\alpha_{(X,\mu)}$, or even α_μ, when the metric d, or the underlying metric space (X,d), is implicit) is defined as

$$\alpha_{(X,d,\mu)}(r) = \sup\big\{1 - \mu(A_r); A \subset X, \mu(A) \geq \tfrac{1}{2}\big\}, \quad r > 0. \tag{1.6}$$

Here $A_r = \{x \in X; d(x,A) < r\}$ is the (open) r-neighborhood of A (with respect to d). A concentration function is less than or equal to $\frac{1}{2}$. When (X,d) is bounded, the enlargements $r > 0$ in (1.6) actually range up to the diameter

$$\mathrm{Diam}(X,d) = \sup\big\{d(x,y); x,y \in X\big\}$$

of (X,d), the concentration function being 0 when r is larger than the diameter. This, however, will not usually be specified. In any case, the concentration function decreases to 0 as $r \to \infty$. Indeed, fix a point x in X. Given $0 < \varepsilon < \frac{1}{2}$, choose r such that the measure of the complement of the ball B with center x and radius r is less than ε. Then, any Borel set A such that $\mu(A) \geq \frac{1}{2}$ intersects B. Hence A_{2r} covers B and thus $1 - \mu(A_{2r}) \leq 1 - \mu(B) < \varepsilon$.

By definition of the concentration function $\alpha_\mu = \alpha_{(X,d,\mu)}$, given a set A with measure $\mu(A) \geq \frac{1}{2}$, the set of points which are within distance $r > 0$ from a point in A has measure larger than or equal to $1 - \alpha_\mu(r)$. If necessary, we agree in the following that $\alpha_\mu(0) = \frac{1}{2}$.

The idea of the concentration of measure phenomenon is that, in a number of basic examples, $\alpha_{(X,d,\mu)}(r)$ decreases rapidly as r, or the dimension of X, is large. In particular, we say that μ has *normal concentration* on (X, d) if there are constants $C, c > 0$ such that, for every $r > 0$,

$$\alpha_{(X,d,\mu)}(r) \leq C\,\mathrm{e}^{-cr^2}. \tag{1.7}$$

As emphasized in Section 1.1, important examples share this normal concentration and we will often be concerned with this property throughout these notes. In particular, as we have seen with (1.2), the normalized invariant measure σ^n on the standard n-sphere $\mathbb{S}^n$, $n \geq 2$, has normal concentration with $c = (n-1)/2$ and $C = 1$, which thus yields strong concentration in high dimension. By (1.4), the canonical Gaussian measure on Euclidean space has this concentration (with $c = \frac{1}{2}$ and $C = 1$). Concentration (1.5) on the cube $\{0,1\}^n$ also belongs to this family. We will also speak of *exponential concentration* if

$$\alpha_{(X,d,\mu)}(r) \leq C\,\mathrm{e}^{-cr}, \quad r > 0.$$

Throughout these notes, we do not pay much attention to sharp constants in normal concentration, although when possible we try to reach the correct exponent $c > 0$. We do not discuss optimal C in normal or exponential concentration.

While the concentration function bounds $1 - \mu(A_r)$ for any measurable set A with $\mu(A) \geq \frac{1}{2}$, it also does so when $\mu(A) \geq \varepsilon > 0$. This is the content of the following easy and useful consequence of the definition of a concentration function.

Lemma 1.1. *Let μ be a probability measure on the Borel sets of a metric space (X, d) with concentration function α_μ. If $\mu(A) \geq \varepsilon > 0$, then*

$$1 - \mu(A_{r_0+r}) \leq \alpha_\mu(r)$$

for any $r > 0$ and $r_0 > 0$ such that $\alpha_\mu(r_0) < \varepsilon$.

Proof. Denote by B the complement of A_{r_0} so that A is included in the complement of B_{r_0}. If $\mu(B) \geq \frac{1}{2}$,

$$\mu(A) \leq 1 - \mu(B_{r_0}) \leq \alpha_\mu(r_0) < \varepsilon$$

which is impossible. Thus $\mu(A_{r_0}) \geq \frac{1}{2}$ so that for every $r > 0$,

$$1 - \mu(A_{r_0+r}) \leq \alpha_\mu(r).$$

The lemma is proved. □

The following simple contraction property shows that concentration functions are decreasing under 1-Lipschitz mappings.

Proposition 1.2. *Let φ be a Lipschitz map between two metric spaces (X, d) and (Y, δ) such that*

$$\delta\big(\varphi(x), \varphi(x')\big) \leq \|\varphi\|_{\mathrm{Lip}} d(x, x') \quad \text{for all } x, x' \in X.$$

Let μ be a probability measure on the Borel sets of X and denote by μ_φ the measure μ pushed forward by φ on the Borel sets of (Y, δ). Then, for every $r > 0$,

$$\alpha_{(Y,\delta,\mu_\varphi)}(r) \leq \alpha_{(X,d,\mu)}\big(r/\|\varphi\|_{\mathrm{Lip}}\big).$$

In particular $\alpha_{\mu_\varphi} \leq \alpha_\mu$ if $\varphi : X \to Y$ is 1-Lipschitz.

For the proof, simply note that if A is a Borel set in Y, then for every $r > 0$,

$$\varphi^{-1}(A_r) \supset \big(\varphi^{-1}(A)\big)_{r/\|\varphi\|_{\mathrm{Lip}}}$$

where the enlargements are understood with respect to d and δ, respectively.

A typical example of application of Proposition 1.2 arises when X is a topological metric group equipped with a (left-) translation invariant metric d and Y is a quotient X/G equipped with the quotient metric

$$\delta(y, y') = \inf\big\{d(x, x'); \varphi(x) = y, \varphi(x') = y'\big\}$$

where $\varphi : X \to X/G$ is the 1-Lipschitz quotient map.

1.3 Deviation inequalities

In this section we discuss equivalent descriptions of concentration properties in terms of deviation and concentration inequalities for Lipschitz functions.

As before, let (X, d) be a metric space with Borel probability measure μ. In the preceding section we defined the concentration function $\alpha_{(X,d,\mu)}(r)$, $r > 0$, as the supremum over all Borel sets A with $\mu(A) \geq \frac{1}{2}$ of $1 - \mu(A_r)$, where we recall that A_r is the enlargement of order r of A in the metric d. One could also define, for every $\varepsilon > 0$,

$$\alpha^{\varepsilon}_{(X,d,\mu)}(r) = \sup\big\{1 - \mu(A_r); A \subset X, \mu(A) \geq \varepsilon\big\}, \quad r > 0,$$

leading essentially to the same concept by Lemma 1.1. The value $\varepsilon = \frac{1}{2}$ is however of particular interest through its connection with medians.

If μ is a probability measure on the Borel sets of (X, d), and if F is a measurable real-valued function on (X, d), we say that m_F is a *median* of F for μ if

$$\mu\big(\{F \leq m_F\}\big) \geq \tfrac{1}{2} \quad \text{and} \quad \mu\big(\{F \geq m_F\}\big) \geq \tfrac{1}{2}\,.$$

A median m_F may not be unique.

For a continuous function F on (X, d), we denote by

$$\omega_F(\eta) = \sup\big\{|F(x) - F(y)|; d(x, y) < \eta\big\}, \quad \eta > 0,$$

its *modulus of continuity*. If m_F is a median of F for μ, and if $A = \{F \leq m_F\}$, note that whenever x is such that $d(x, y) < \eta$ for some $y \in A$, then

$$F(x) \leq F(y) + \omega_F(\eta) \leq m_F + \omega_F(\eta).$$

Hence, since $\mu(A) \geq \frac{1}{2}$, by definition of the concentration function $\alpha_\mu = \alpha_{(X,d,\mu)}$,

$$\mu\big(\{F > m_F + \omega_F(\eta)\}\big) \leq \alpha_\mu(\eta). \tag{1.8}$$

Similarly with $A = \{F \geq m_F\}$,

$$\mu\big(\{F < m_F - \omega_F(\eta)\}\big) \leq \alpha_\mu(\eta).$$

(Alternatively, replace F by $-F$.) In particular,

$$\mu\big(\{|F - m_F| > \omega_F(\eta)\}\big) \leq 2\alpha_\mu(\eta). \tag{1.9}$$

Inequalities (1.8) and (1.9) on the n-dimensional sphere $\mathbb{S}^n$ are sometimes called *Lévy's inequalities.* The high dimensional effect on $\mathbb{S}^n$ is expressed by the fact that $\alpha_{\sigma^n}(\eta)$ is small when n is large, showing thus that functions on $\mathbb{S}^n$ with small local oscillation are nearly constant on almost all of the space (with respect to σ^n).

It will be more convenient in the sequel to work with Lipschitz functions and Lipschitz coefficients rather than with moduli of continuity. Let us detail again (1.8) and (1.9) on Lipschitz functions. A real-valued function F on (X, d) is said to be *Lipschitz* if

$$\|F\|_{\mathrm{Lip}} = \sup_{x \neq y} \frac{|F(x) - F(y)|}{d(x,y)} < \infty.$$

Clearly $\omega_F(\eta) \leq \eta \|F\|_{\mathrm{Lip}}$ for every $\eta > 0$. We say that F is 1-Lipschitz if $\|F\|_{\mathrm{Lip}} \leq 1$. The class of Lipschitz functions is stable under the operations min and max.

If F is Lipschitz on (X, d) and if $A = \{F \leq m\}$, then, for every $r > 0$, $A_r \subset \{F < m + r\|F\|_{\mathrm{Lip}}\}$. Therefore, if $m = m_F$ is a median of F for μ, we get as for (1.8) that for every $r > 0$,

$$\mu\big(\{F \geq m_F + r\}\big) \leq \alpha_\mu\big(r/\|F\|_{\mathrm{Lip}}\big). \tag{1.10}$$

We speak of (1.10) (and (1.8)) as a *deviation inequality.*

By Lemma 1.1, if $\mu(\{F \leq m\}) \geq \varepsilon > 0$, then for every $r > 0$,

$$\mu\big(\{F \geq m + r_0 + r\}\big) \leq \alpha_\mu\big(r/\|F\|_{\mathrm{Lip}}\big) \tag{1.11}$$

where $\alpha_\mu(r_0/\|F\|_{\mathrm{Lip}}) < \varepsilon$. Similarly, the inequality (1.10) would hold with $\alpha^\varepsilon_{(X,d,\mu)}$ as soon as m_F is such that $\mu(\{F \leq m_F\}) \geq \varepsilon > 0$. But, as before, the particular choice of $\varepsilon = \frac{1}{2}$ allows us to repeat the same argument with $-F$ to get that

$$\mu\big(\{F \leq m_F - r\}\big) \leq \alpha_\mu\big(r/\|F\|_{\mathrm{Lip}}\big). \tag{1.12}$$

Therefore, together with (1.10), we deduce that for every $r > 0$,

$$\mu\big(\{|F - m_F| \geq r\}\big) \leq 2\alpha_\mu\big(r/\|F\|_{\mathrm{Lip}}\big). \tag{1.13}$$

This inequality (as well as (1.9)) describes a *concentration inequality* of F around its median (one of them) with rate α_μ. According to the relative size of α_μ and $\|F\|_{\mathrm{Lip}}$, the Lipschitz function F "concentrates" around one constant value on a portion of the space of large measure. Moreover, it should already be emphasized that m_F and $\|F\|_{\mathrm{Lip}}$ might be of rather different scales, an observation of fundamental importance in applications. On the other hand, concentration usually does not yield any particular kind of information on the size of the Lipschitz functions themselves (in particular of m_F).

By homogeneity, it is enough to consider the preceding deviation and concentration inequalities for 1-Lipschitz functions.

The deviation or concentration inequalities on Lipschitz functions (1.10) and (1.13) are actually equivalent to the corresponding statement on sets. Let A be a Borel set in (X, d) with $\mu(A) \geq \frac{1}{2}$. Set $F(x) = d(x, A)$, $x \in X$. Clearly $\|F\|_{\mathrm{Lip}} \leq 1$ while

$$\mu(\{F = 0\}) \geq \mu(A) \geq \frac{1}{2}\,.$$

Hence, since $F \geq 0$, 0 is a median of F for μ, and thus, by (1.10), for every $r > 0$,

$$1 - \mu(A_r) = \mu(\{F \geq r\}) \leq \alpha_\mu(r).$$

We may summarize these conclusions in a statement.

Proposition 1.3. *Let μ be a Borel probability measure on a metric space (X, d). Let F be a real-valued continuous function on (X, d) with modulus of continuity ω_F and let m_F be a median of F for μ. Then, for every $\eta > 0$,*

$$\mu\big(\{F > m_F + \omega_F(\eta)\}\big) \leq \alpha_\mu(\eta).$$

In particular, if F is Lipschitz, for any $r > 0$,

$$\mu(\{F \geq m_F + r\}) \leq \alpha_\mu(r/\|F\|_{\mathrm{Lip}})$$

and

$$\mu(\{|F - m_F| \geq r\}) \leq 2\alpha_\mu(r/\|F\|_{\mathrm{Lip}}).$$

Conversely, if for some non-negative function α on $\mathbb{R}_+$,

$$\mu(\{F \geq m_F + r\}) \leq \alpha(r)$$

for any 1-Lipschitz function F with median m_F and any $r > 0$, then $\alpha_\mu \leq \alpha$.

The previous proposition has the following interesting consequences.

Corollary 1.4. *If μ on (X, d) has concentration function $\alpha_\mu = \alpha_{(X,d,\mu)}$, for any two non-empty Borel sets A and B in X,*

$$\mu(A)\mu(B) \leq 4\alpha_\mu\big(d(A, B)/2\big)$$

where $d(A, B) = \inf\{d(x, y); x \in A, y \in B\}$.

Proof. Let $2r = d(A, B) > 0$. Consider the 1-Lipschitz function $F(y) = d(y, B)$ and denote by m_F a median of F for μ. Since $F = 0$ on B and $F \geq 2r$ on A,

$$\begin{aligned}
\mu(A)\mu(B) &\leq \mu \otimes \mu\Big(\big\{(x, y); \big|F(x) - F(y)\big| \geq 2r\big\}\Big) \\
&\leq 2\mu(\{|F - m_F| \geq r\}) \\
&\leq 4\alpha_\mu(r),
\end{aligned}$$

which is the desired result. □

Corollary 1.5. *If μ on (X,d) has concentration function $\alpha_\mu = \alpha_{(X,d,\mu)}$, for any 1-Lipschitz function F on (X,d) and any $r>0$,*

$$\mu\otimes\mu\Big(\big\{(x,y)\in X\times X; \big|F(x)-F(y)\big|\geq r\Big) \leq 2\alpha_\mu\Big(\frac{r}{2}\Big).$$

Conversely, if for some non-negative function α on $\mathbb{R}_+$, all 1-Lipschitz functions F and all $r>0$,

$$\mu\otimes\mu\Big(\big\{(x,y)\in X\times X; \big|F(x)-F(y)\big|\geq r\Big) \leq \alpha(r),$$

then $\alpha_\mu \leq 2\alpha$.

Proof. The first assertion follows from the fact, already used in Corollary 1.4, that

$$\mu\otimes\mu\Big(\big\{(x,y); \big|F(x)-F(y)\big|\geq 2r\big\}\Big) \leq 2\mu\big(\{|F-m_F|\geq r\}\big).$$

Conversely, take A with $\mu(A)\geq \frac{1}{2}$. Applying the hypothesis to $F(x)=d(x,A)$, $x\in X$, we get as in Corollary 1.4 that for every $r>0$,

$$\begin{aligned}\mu(A)\big(1-\mu(A_r)\big) &\leq \mu\otimes\mu\Big(\big\{(x,y); \big|F(x)-F(y)\big|\geq r\big\}\Big)\\ &\leq \alpha(r)\end{aligned}$$

from which the desired claim follows. □

A particular situation for deviation inequalities under some level occurs for convex functions and the next statement is a short digression on this theme. Assume for simplicity that $X=\mathbb{R}^n$ equipped with its standard Euclidean metric $|\cdot|$ and denote by ∇F the gradient of a smooth function F on $\mathbb{R}^n$.

Proposition 1.6. *Let μ be a probability measure on the Borel sets of $\mathbb{R}^n$, and let F be smooth and convex on $\mathbb{R}^n$ such that for some $m\in\mathbb{R}$ and $L>0$,*

$$\mu\big(\{F\geq m; |\nabla F|\leq L\}\big)\geq \varepsilon>0.$$

Then, for every $r>0$,

$$\mu\big(\{F\leq m-L(r_0+r)\}\big)\leq \alpha_\mu(r)$$

where $\alpha_\mu(r_0)<\varepsilon$.

Proof. Applying Lemma 1.1 to $r_1=r_0+r$, it is enough to show that whenever $A=\{F\geq m, |\nabla F|\leq L\}$, then

$$A_{r_1}\subset\{F>m-Lr_1\}.$$

But since F is (smooth and) convex, for any $x,y\in\mathbb{R}^n$,

$$F(y)\leq F(x)+\nabla F(y)\cdot(y-x).$$

Hence, if $y\in A$ and $|x-y|<r_1$,

$$F(x)\geq F(y)-\big|\nabla F(y)\big||x-y|>m-Lr_1.$$

The proposition is proved. □

When m is a median of F in Proposition 1.6, it is usually impossible to expect that $\mu(A) \geq \frac{1}{2}$ where

$$A = \{F \geq m, |\nabla F| \leq L\}.$$

To estimate $\mu(A)$ from below we may however write

$$\mu(A) \geq \mu(\{F \geq m\}) - \mu(\{|\nabla F| > L\}). \tag{1.14}$$

If F is Lipschitz on $\mathbb{R}^n$, by Rademacher's theorem, F is almost everywhere differentiable and $\|\nabla F\|_\infty = \|F\|_{\mathrm{Lip}}$. Proposition 1.6 thus shows that for Lipschitz convex functions, the deviation inequalities under some level m are governed, using (1.14), by the L^0-norm of the gradient of F rather than by its L^∞-norm as in (1.12). This is a useful observation in applications. Similar conclusions of course apply to deviation inequalities above some level for concave functions.

Inequality (1.13) describes a concentration property of the Lipschitz function F around some median value m_F. The median m_F may actually be replaced by the mean of F. We first show a converse result in this direction.

Proposition 1.7. *Let μ be a Borel probability measure on a metric space (X,d). Assume that for some non-negative function α on $\mathbb{R}_+$ and any bounded 1-Lipschitz function F on (X,d),*

$$\mu\big(\{F \geq \textstyle\int F d\mu + r\}\big) \leq \alpha(r) \tag{1.15}$$

for every $r > 0$. Then

$$1 - \mu(A_r) \leq \alpha\big(\mu(A)\, r\big)$$

for every Borel set A with $\mu(A) > 0$ and every $r > 0$. In particular,

$$\alpha_{(X,d,\mu)}(r) \leq \alpha\big(\tfrac{r}{2}\big), \quad r > 0.$$

Moreover, if α is such that $\lim_{r\to\infty} \alpha(r) = 0$, any 1-Lipschitz function F is integrable with respect to μ and, provided α is continuous, satisfies (1.15).

Proof. Take A with $\mu(A) > 0$ and fix $r > 0$. Consider $F(x) = \min(d(x,A), r)$, $x \in X$. Clearly $\|F\|_{\mathrm{Lip}} \leq 1$ while

$$\int F d\mu \leq \big(1 - \mu(A)\big) r.$$

By the hypothesis,

$$\begin{aligned} 1 - \mu(A_r) &= \mu\big(\{F \geq r\}\big) \\ &\leq \mu\big(\{F \geq \textstyle\int F d\mu + \mu(A)\, r\}\big) \\ &\leq \alpha\big(\mu(A)\, r\big). \end{aligned}$$

In particular, if $\mu(A) \geq \frac{1}{2}$, $1 - \mu(A_r) \leq \alpha(\frac{r}{2})$ so that the first claim follows.

Now let F be a 1-Lipschitz function on (X,d). For every $n \geq 0$, $F_n = \min(|F|, n)$ is again 1-Lipschitz and bounded. Applying (1.15) to $-F_n$, for every $r > 0$,

$$\mu\big(\{F_n \leq \textstyle\int F_n d\mu - r\}\big) \leq \alpha(r). \tag{1.16}$$

Choose m such that $\mu(\{|F| \leq m\}) \geq \frac{1}{2}$ and r_0 such that $\alpha(r_0) < \frac{1}{2}$. Since for every n, $\mu(\{F_n \leq m\}) \geq \frac{1}{2}$, intersecting with (1.16) for $r = r_0$, we get that, independently of n,

$$\int F_n d\mu \leq m + r_0$$

and thus $\int |F| d\mu < \infty$ by monotone convergence. Then we apply (1.15) to $\min(\max(F, -n), n)$ and let $n \to \infty$. Proposition 1.7 is established. □

In Proposition 1.7, we lose the factor 2 in the concentration function. This may be improved by an iteration of the argument. In the same spirit, observe that (1.15) together with the same inequality for $-F$ contains a concentration inequality around the mean

$$\mu\Big(\Big\{\big|F - \int F d\mu\big| \geq r\Big\}\Big) \leq 2\alpha(r), \quad r > 0. \tag{1.17}$$

Now, if r_0 is chosen so that $2\alpha(r_0) < \frac{1}{2}$, any median m_F of F is such that

$$\Big|m_F - \int F d\mu\Big| \leq r_0.$$

From this together with (1.17) it follows that

$$\mu\big(\{|F - m_F| \geq r + r_0\}\big) \leq 2\alpha(r), \quad r > 0. \tag{1.18}$$

In the case of normal concentration $\alpha(r) = C\,\mathrm{e}^{-cr^2}$, for example, we then get via Proposition 1.3 a concentration function α_μ asymptotically of the same order as $r \to \infty$, namely

$$\alpha_\mu(r) \leq C'\,\mathrm{e}^{-cr^2 + c'r}, \quad r > 0.$$

We will not be concerned in this work with sharp constants in concentration functions, and usually present bounds on concentration functions using the next simple Proposition 1.7 with factor $\frac{1}{2}$. Moreover, in normal concentration functions $\alpha(r) = C\,\mathrm{e}^{-cr^2}$, we will never describe sharp values of C. However, when the sharp concentration inequalities for Lipschitz functions around the median or the mean are available, we present them simultaneously. As we will see in the next chapter, isoperimetric inequalities do usually provide optimal concentration functions.

The next proposition formalizes the argument leading to (1.18) with the mean replaced by any constant value. To this task, we may actually work at the level of one single function F.

Proposition 1.8. *Let F be a measurable function on a probability space $(X, \mathcal{A}, \mu)$. Assume that for some $a_F \in \mathbb{R}$ and a non-negative function α on $\mathbb{R}_+$ such that $\lim_{r\to\infty} \alpha(r) = 0$,*

$$\mu\big(\{|F - a_F| \geq r\}\big) \leq \alpha(r)$$

for all $r > 0$. Then

$$\mu\big(\{|F - m_F| \geq r + r_0\}\big) \leq \alpha(r), \quad r > 0,$$

where m_F is a median of F for μ and where $r_0 > 0$ is such that $\alpha(r_0) < \frac{1}{2}$. If moreover $\overline{\alpha} = \int_0^\infty \alpha(r)dr < \infty$, then F is integrable, $|a_F - \int F d\mu| \leq \overline{\alpha}$ and for every $r > 0$,

$$\mu\Big(\Big\{\big|F - \int F d\mu\big| \geq r + \overline{\alpha}\Big\}\Big) \leq \alpha(r).$$

In particular, if $\alpha(r) \leq C\,\mathrm{e}^{-cr^p}$, $0 < p < \infty$, $r > 0$, *then*

$$\mu\big(\{|F - M| \geq r\}\big) \leq C'\mathrm{e}^{-\kappa_p cr^p}, \quad r > 0,$$

where M *is either the mean or a median of* F *for* μ *and where* $C' > 0$ *only depends on* C *and* p *and* $\kappa_p > 0$ *only depends on* p.

Proof. The first part follows from the argument leading to (1.18). For the second part, note that

$$\int |F - a_F| d\mu = \int_0^\infty \mu\big(\{|F - a_F| \geq r\}\big)dr \leq \overline{\alpha}.$$

Therefore $\int |F|d\mu < \infty$ and $|a_F - \int F d\mu| \leq \overline{\alpha}$ from which the desired claim easily follows. If $\alpha(r) \leq C\,\mathrm{e}^{-cr^p}$, $0 < p < \infty$, $r > 0$, we may choose $r_0 > 0$ such that $\mathrm{e}^{cr_0^p} = 2C$ so that when $r \leq r_0$,

$$\mu\big(\{|F - m_F| \geq r\}\big) \leq 1 \leq 2C\,\mathrm{e}^{-cr_0^p} \leq 2C\,\mathrm{e}^{-cr^p}$$

while when $r \geq r_0$,

$$\mu\big(\{|F - m_F| \geq r\}\big) \leq C\,\mathrm{e}^{-c(r-r_0)^p} \leq C'\mathrm{e}^{-\kappa_p cr^p}.$$

In the same way,

$$\overline{\alpha} \leq \int_0^\infty C\,\mathrm{e}^{-cr^p} dr = K_p C c^{-1/p}$$

where $K_p > 0$ only depends on p. A similar argument, in accordance with $r \leq \overline{\alpha}$ or $r \geq \overline{\alpha}$, yields the concentration inequality around the mean of F. □

Normal concentration implies strong integrability properties of Lipschitz functions. This is the content of the simple proposition that immediately follows by integration in $r > 0$.

Proposition 1.9. *Let* F *be a measurable function on some probability space* $(X, \mathcal{A}, \mu)$ *such that for some* $a_F \in \mathbb{R}$ *and some constants* $C, c > 0$,

$$\mu\big(\{|F - a_F| \geq r\}\big) \leq C\,\mathrm{e}^{-cr^2}$$

for every $r > 0$. *Then*

$$\int \mathrm{e}^{\rho F^2} d\mu < \infty$$

for every $\rho < c$. *Furthermore,*

$$\left| \int F d\mu - a_F \right| \leq \frac{C}{2}\sqrt{\frac{\pi}{c}} \quad \text{and} \quad \mathrm{Var}_\mu(F) \leq \frac{C}{c}\,.$$

Proof. From the hypothesis, for every $r > |a_F|$,

$$\mu\big(\{|F| \geq r\}\big) \leq \mu\big(\{|F - a_F| \geq r - |a_F|\}\big) \leq C\,\mathrm{e}^{-c(r-|a_F|)^2}.$$

Now, by Fubini's theorem,

$$\begin{aligned}\int \mathrm{e}^{\rho F^2} d\mu &= 1 + \int_0^\infty 2\rho r \mu\big(\{|F| \geq r\}\big)\mathrm{e}^{\rho r^2} dr \\ &\leq \mathrm{e}^{\rho a_F^2} + \int_{|a_F|}^\infty 2\rho r \mu\big(\{|F| \geq r\}\big)\, \mathrm{e}^{\rho r^2} dr \\ &\leq \mathrm{e}^{\rho a_F^2} + \int_{|a_F|}^\infty 2C\rho r\, \mathrm{e}^{-c(r-|a_F|)^2}\, \mathrm{e}^{\rho r^2} dr\end{aligned}$$

from which the first claim follows. As in the proof of Proposition 1.8,

$$\begin{aligned}\left|\int F d\mu - a_F\right| &\leq \int |F - a_F| d\mu \\ &= \int_0^\infty \mu\big(\{|F - a_F| \geq r\}\big) dr \\ &\leq \int_0^\infty C\mathrm{e}^{-cr^2} dr = \frac{C}{2}\sqrt{\frac{\pi}{c}}\end{aligned}$$

while

$$\begin{aligned}\mathrm{Var}_\mu(F) &\leq \int |F - a_F|^2 d\mu \\ &= \int_0^\infty 2r\mu\big(\{|F - a_F| \geq r\}\big) dr \\ &\leq \int_0^\infty 2Cr\mathrm{e}^{-cr^2} dr = \frac{C}{c}\,.\end{aligned}$$

□

The preceding statement is typical of one basic aspect of concentration in that it does not tell anything about the size of F but only bounds discrepancies around a mean value.

The following proposition is a further description of normal concentration.

Proposition 1.10. *Let μ be a Borel probability measure on a metric space (X, d). Then (X, d, μ) has normal concentration $\alpha_\mu(r) \leq C\,\mathrm{e}^{-cr^2}$, $r > 0$, if and only if there is a constant $K > 0$ (depending only on C) such that for every $q \geq 1$ and every 1-Lipschitz function F on (X, d),*

$$\left\|F - \textstyle\int F d\mu\right\|_q \leq K\sqrt{\frac{q}{c}}$$

where $\|\cdot\|_q$ is the L^q-norm with respect to μ.

Proof. If $\alpha_\mu(r) \leq C\,\mathrm{e}^{-cr^2}$, $r \geq 0$, we know from Proposition 1.8 that for every 1-Lipschitz function F, and every $r > 0$,

$$\mu\big(\{|F - \textstyle\int F d\mu| \geq r\}\big) \leq C'\,\mathrm{e}^{-\kappa cr^2}$$

where $C' > 0$ only depends on $C > 0$ and $\kappa > 0$ is numerical. Then, as before, for $q \geq 1$,

$$\begin{aligned}\left\|F - \textstyle\int F d\mu\right\|_q^q &= \int_0^\infty q r^{q-1} \mu\big(\{|F - \textstyle\int F d\mu| \geq r\}\big) dr \\ &\leq \int_0^\infty 2C' q r^{q-1} \mathrm{e}^{-\kappa cr^2} dr\end{aligned}$$

from which the implication follows since

$$\int_0^\infty r^{q-1}\mathrm{e}^{-\kappa c r^2}dr \sim \left(\frac{q}{c}\right)^{q/2}$$

as $q \to \infty$. Conversely, by Chebyshev's inequality, for every $r > 0$ and $q \geq 1$,

$$\mu\big(\{|F - \textstyle\int F d\mu| \geq r\}\big) \leq K^q\left(\frac{q}{c}\right)^{q/2} r^{-q}$$

from which normal concentration (with exponential rate proportional to c) follows by optimization in $q \geq 1$. (The same description of course holds with the median instead of the mean.) □

As a consequence of Proposition 1.10, if (X, d, μ) has normal concentration, there exists $K > 0$ such that for every $q \geq 1$ and every Lipschitz function F on (X, d),

$$\|F\|_q \leq \|F\|_1 + K\sqrt{q}\,\|F\|_{\mathrm{Lip}}. \tag{1.19}$$

Propositions 1.9 and 1.10 clearly extend to concentration functions of the type $C\,\mathrm{e}^{-cr^p}$, $p > 0$ (or more general sufficiently small concentration functions). The growth rate in $q \geq 1$ is then $q^{1/p}$ in Proposition 1.10.

Proposition 1.7 is a convenient tool to handle concentration in product spaces with which we conclude this section. If (X, d) and (Y, δ) are metric spaces, we equip the Cartesian product space $X \times Y$ with the ℓ^1-metric

$$d(x, x') + \delta(y, y'), \quad x, x' \in X,\ y, y' \in Y. \tag{1.20}$$

Proposition 1.11. *Let μ and ν be Borel probability measures on metric spaces (X, d) and (Y, δ) respectively. Let α and β be non-negative functions on $\mathbb{R}_+$ such that whenever $F : X \to \mathbb{R}$ and $G : Y \to \mathbb{R}$ are bounded and 1-Lipschitz on their respective spaces, then for every $r > 0$,*

$$\mu\big(\{F \geq \textstyle\int F d\mu + r\}\big) \leq \alpha(r)$$

and

$$\nu\big(\{G \geq \textstyle\int G d\nu + r\}\big) \leq \beta(r).$$

Then, if $\mu \otimes \nu$ is the product measure of μ and ν on $X \times Y$ equipped with the ℓ^1-metric (1.20), for any bounded 1-Lipschitz function F on the product space $X \times Y$ and any $r > 0$,

$$\mu \otimes \nu\big(\{F \geq \textstyle\int F d\mu \otimes \nu + 2r\}\big) \leq \alpha(r) + \beta(r).$$

Proof. Set, for every $x \in X$, $y \in Y$, $F^y(x) = F(x, y)$ and $G(y) = \int F^y d\mu$, and observe that F^y and G are 1-Lipschitz on their respective spaces. Therefore, for every $r > 0$,

$$\begin{aligned}
&\mu \otimes \nu\big(\{F \geq \textstyle\int F d\mu \otimes \nu + 2r\}\big) \\
&\qquad \leq \mu \otimes \nu\big(\big\{(x, y) \in X \times Y; F^y(x) \geq \textstyle\int F^y d\mu + r\big\}\big) \\
&\qquad\quad + \nu\big(\big\{G \geq \textstyle\int G d\nu + r\big\}\big) \\
&\qquad \leq \alpha(r) + \beta(r).
\end{aligned}$$

The proposition is established. □

While Proposition 1.11 describes concentration results in product spaces, these are not well suited to concentration bounds which are independent of the number of spaces in the product (dimension free concentration). It will be one main task addressed in Chapters 4, 5 and 6 to develop tools to reach such dimension free bounds of fundamental importance in applications.

1.4 Observable diameter

The notion of observable diameter is somewhat dual to the one of concentration function. It describes the diameter of a metric space (X, d) viewed through a given probability measure μ on the Borel sets of (X, d).

We fix $\kappa > 0$ to be thought of as small ($\kappa = 10^{-10}$ in [Grom2]!). According to M. Gromov [Grom2], define first the partial diameter $\mathrm{PartDiam}_\mu(X, d)$ of (X, d) with respect to μ as the infimal D such that there exists a subset A of X with diameter less than or equal to D and measure $\mu(A) \geq 1-\kappa$. This diameter is clearly monotone for the Lipschitz ordering: if $\varphi : (X, d) \to (Y, \delta)$ is 1-Lipschitz, and if μ_φ is the pushed forward measure μ by φ, then $\mathrm{PartDiam}_{\mu_\varphi}(Y, \delta) \leq \mathrm{PartDiam}_\mu(X, d)$ (for all $\kappa > 0$). What is not obvious is that the partial diameter may dramatically decrease under all 1-Lipschitz maps from X to a certain Y, which we always take to be $\mathbb{R}$. We then define the *observable diameter* $\mathrm{ObsDiam}_\mu(X, d)$ of (X, d) with respect to μ as the supremum of $\mathrm{PartDiam}_{\mu_F}(\mathbb{R})$ over all image measures μ_F of μ by a 1-Lipschiz map $F : X \to \mathbb{R}$.

Following [Grom2], we think of μ as a state on the configuration space (X, d) and a Lipschitz map $F : X \to \mathbb{R}$ is interpreted as an observable giving the tomographic image μ_F on $\mathbb{R}$. We watch μ_F and can only distinguish a part of its support of measure $1 - \kappa$.

The next simple statement connects the observable diameter $\mathrm{ObsDiam}_\mu(X, d)$ with the concentration function $\alpha_\mu = \alpha_{(X,d,\mu)}$. Let α_μ^{-1} be the generalized inverse function of the non-increasing function α_μ, that is,

$$\alpha_\mu^{-1}(\varepsilon) = \inf\big\{r > 0; \alpha_\mu(r) \leq \varepsilon\big\}, \quad \varepsilon > 0.$$

Proposition 1.12. *Let (X, d, μ) be a metric measure space, and let $\kappa > 0$. Then*

$$\mathrm{ObsDiam}_\mu(X, d) \leq 2\alpha_\mu^{-1}\Big(\frac{\kappa}{2}\Big).$$

Proof. It is an easy consequence of the concentration inequalities for Lipschitz functions of the preceding sections. Indeed, if $F : X \to \mathbb{R}$ is 1-Lipschitz, we know from (1.13) that for every $r > 0$,

$$\mu\big(\{|F - m_F| \geq r\}\big) \leq 2\alpha_\mu(r)$$

where m_F is a median of F for μ. Hence, if μ_F denotes the image measure of μ by F, the interval $]m_F - r, m_F + r[$ has length $2r$ and μ_F-measure larger than or equal to $1 - 2\alpha_\mu(r)$. The proposition then follows from the definition of the inverse function α_μ^{-1} of α_μ. □

As an example, if μ has normal concentration $\alpha_\mu(r) \leq C\,\mathrm{e}^{-cr^2}$, $r > 0$, on (X,d), then

$$\mathrm{ObsDiam}_\mu(X,d) \leq 2\sqrt{\frac{1}{c}\,\log\frac{2C}{\kappa}}\,. \tag{1.21}$$

The important parameter in (1.21) is the rate c in the exponential decay of the concentration function, the value of $C > 0$ being usually a numerical constant that simply modifies the numerical value of κ by a factor. For example, by (1.2), (1.21) for the standard n-sphere $\mathbb{S}^n$ may loosely be described by saying that

$$\mathrm{ObsDiam}_{\sigma^n}(\mathbb{S}^n) = O\Big(\frac{1}{\sqrt{n}}\Big)$$

as n is large, which is of course in strong contrast with the diameter of $\mathbb{S}^n$ itself as a metric space. Similarly, the observable diameter of Euclidean space with respect to Gaussian measures is bounded.

1.5 Expansion coefficient

Expansion coefficients are a natural multiplicative description of exponential concentration.

As before, let μ be a probability measure on the Borel sets of a metric space (X,d). Define the *expansion coefficient* of μ on (X,d) of order $\varepsilon > 0$ as

$$\mathrm{Exp}_\mu(\varepsilon) = \inf\big\{e \geq 1; \mu(B_\varepsilon) \geq e\mu(B), B \subset X, \mu(B_\varepsilon) \leq \tfrac{1}{2}\big\}$$

where we recall that B_ε is the (open) ε-neighborhood of B with respect to d.

The definition of $\mathrm{Exp}_\mu(\varepsilon)$ shows that whenever B is such that $\mu(B_{k\varepsilon}) \leq \frac{1}{2}$ for some integer $k \geq 1$, then

$$\mu(B) \leq \mathrm{Exp}_\mu(\varepsilon)^{-k}\mu(B_{k\varepsilon}) \leq \frac{1}{2}\,\mathrm{Exp}_\mu(\varepsilon)^{-k}. \tag{1.22}$$

In particular, if $\mathrm{Exp}_\mu(\varepsilon) > 1$, B has very small measure. Thinking of B as the complement of some large r-neighborhood of a set A with measure $\mu(A) \geq \frac{1}{2}$ immediately leads to the exponential decay of the concentration function α_μ of μ on (X,d).

Proposition 1.13. *If for some $\varepsilon > 0$, $\mathrm{Exp}_\mu(\varepsilon) \geq e > 1$, then (X,d,μ) has exponential concentration*

$$\alpha_\mu(r) \leq \frac{e}{2}\,\mathrm{e}^{-r(\log e)/\varepsilon}, \quad r > 0.$$

Proof. Let A be a Borel set in (X,d) with $\mu(A) \geq \frac{1}{2}$. If B is the complement of $A_{k\varepsilon}$, then $B_{k\varepsilon}$ is contained in the complement of A and thus $\mu(B_{k\varepsilon}) \leq \frac{1}{2}$. Therefore, by (1.22),

$$1 - \mu(A_{k\varepsilon}) \leq \frac{1}{2e^k}\,.$$

We then simply interpolate between $k\varepsilon$ and $(k+1)\varepsilon$ to get the desired result. □

1.6 Laplace bounds and infimum-convolutions

In this section, we provide some simple and useful tools to establish concentration properties, either through exponential deviation inequalities for Lipschitz functions or by infimum-convolution arguments.

Let (X, d) be a metric space and let μ be a probability measure on the Borel sets of (X, d). Define, for $\lambda \geq 0$, the *Laplace functional* of μ on (X, d) as

$$\mathrm{E}_{(X,d,\mu)}(\lambda) = \sup \int \mathrm{e}^{\lambda F} d\mu$$

where the supremum runs over all (bounded) mean zero 1-Lipschitz functions F on (X, d). We often write more simply $\mathrm{E}_\mu = \mathrm{E}_{(X,d,\mu)}$.

The following elementary proposition bounds the concentration function $\alpha_{(X,d,\mu)}$ of μ on (X, d) by $\mathrm{E}_{(X,d,\mu)}$.

Proposition 1.14. *Under the preceding notation,*

$$\alpha_{(X,d,\mu)}(r) \leq \inf_{\lambda \geq 0} \mathrm{e}^{-\lambda r/2}\, \mathrm{E}_{(X,d,\mu)}(\lambda), \quad r > 0.$$

In particular, if

$$\mathrm{E}_{(X,d,\mu)}(\lambda) \leq \mathrm{e}^{\lambda^2/2c}, \quad \lambda \geq 0,$$

then, every 1-Lipschitz function $F : X \to \mathbb{R}$ is integrable and for every $r \geq 0$,

$$\mu\big(\{F \geq \textstyle\int F d\mu + r\}\big) \leq \mathrm{e}^{-cr^2/2}$$

and (X, d, μ) has normal concentration

$$\alpha_{(X,d,\mu)}(r) \leq \mathrm{e}^{-cr^2/8}, \quad r > 0.$$

If $\mathrm{E}_{(X,d,\mu)}(\lambda_0) < \infty$ for some $\lambda_0 > 0$, then (X, d, μ) has exponential concentration.

For the proof, simply note that by Chebyshev's exponential inequality, for any mean zero 1-Lipschitz function $F : X \to \mathbb{R}$, and any r and $\lambda > 0$,

$$\mu\big(\{F \geq r\}\big) \leq \mathrm{e}^{-\lambda r}\, \mathrm{E}_{(X,d,\mu)}(\lambda).$$

Optimizing in λ, the various conclusions immediately follow from Proposition 1.7.

As for concentration functions, the Laplace functional is decreasing under 1-Lipschitz maps. Moreover, the Laplace functional $\mathrm{E}_{(X,d,\mu)}$ is a convenient tool to handle concentration in product spaces with respect to the ℓ^1-metric. If (X, d) and (Y, δ) are two metric spaces, we equip the product space $X \times Y$ with the metric (1.20),

$$d(x, x') + \delta(y, y'), \quad x, x' \in X,\ y, y' \in Y.$$

Proposition 1.15. *Under the preceding notation,*

$$\mathrm{E}_{(X\times Y, d+\delta, \mu\otimes\nu)} \leq \mathrm{E}_{(X,d,\mu)}\, \mathrm{E}_{(Y,\delta,\nu)}.$$

Proof. It is similar to the proof of Proposition 1.11. Let F be a mean zero 1-Lipschitz function on the product space. Set, for every $x \in X$, $y \in Y$, $F^y(x) = F(x,y)$ and $G(y) = \int F^y d\mu$. Observe that F^y and G are 1-Lipschitz on their respective spaces. We then write, for every $\lambda \geq 0$,

$$\begin{aligned}\int e^{\lambda F} d\mu \otimes \nu &= \int e^{\lambda G(y)} \Big(\int e^{\lambda [F^y(x) - \int F^y d\mu]} d\mu(x) \Big) d\nu(y) \\ &\leq E_{(X,d,\mu)}(\lambda) \int e^{\lambda G} d\nu\end{aligned}$$

from which the claim follows since $\int G d\nu = 0$. □

The next statement is a simple illustration of the use of Laplace functionals. It will describe a basic concentration behavior in product spaces with respect to the ℓ^1-metric.

Proposition 1.16. *If* $\mathrm{Diam}\,(X,d) = D < \infty$, *then, for any probability measure* μ *on the Borel sets of* (X,d),

$$E_{(X,d,\mu)}(\lambda) \leq e^{D^2\lambda^2/2}, \quad \lambda \geq 0.$$

Proof. Let F be a mean zero 1-Lipschitz function on (X,d). By Jensen's inequality, for every $\lambda \geq 0$,

$$\begin{aligned}\int e^{\lambda F} d\mu &\leq \iint e^{\lambda [F(x) - F(y)]} d\mu(x) d\mu(y) \\ &\leq \sum_{i=0}^{\infty} \frac{(D\lambda)^{2i}}{(2i)!} \\ &\leq e^{D^2\lambda^2/2},\end{aligned}$$

and the proposition is proved. □

As a consequence of Propositions 1.14, 1.15 and 1.16, we get the following simple, yet important, corollary.

Corollary 1.17. *Let* $P = \mu_1 \otimes \cdots \otimes \mu_n$ *be any product probability measure on the Cartesian product* $X = X_1 \times \cdots \times X_n$ *of metric spaces* (X_i, d_i) *with finite diameters* D_i, $i = 1, \dots, n$, *equipped with the* ℓ^1*-metric* $d = \sum_{i=1}^n d_i$. *Then, if* F *is a 1-Lipschitz function on* (X,d), *for every* $r \geq 0$,

$$P\big(\{F \geq \textstyle\int F dP + r\}\big) \leq e^{-r^2/2D^2}$$

where $D^2 = \sum_{i=1}^n D_i^2$. *In particular,*

$$\alpha_P(r) \leq e^{-r^2/8D^2}, \quad r > 0.$$

Applied to sums $S = Y_1 + \cdots + Y_n$ of real-valued independent random variables $Y_1, \dots, Y_n$ on some probability space $(\Omega, \mathcal{A}, \mathbb{P})$ such that $u_i \leq Y_i \leq v_i$, $i = 1, \dots, n$, Corollary 1.17 yields (with worse constants) a Hoeffding type inequality [Hoe] (see [Sto], [MD2])

$$\mathbb{P}\big(\{S \geq \mathbb{E}(S) + r\}\big) \leq e^{-r^2/2D^2} \tag{1.23}$$

for every $r \geq 0$ where $D^2 \geq \sum_{i=1}^n (v_i - u_i)^2$. Exponential inequalities such as (1.23), obtained from Laplace transform estimates, have actually a long run in the study of limit theorems in classical probability theory going back to S. Bernstein, A. Kolmogorov, Y. Prokhorov etc (cf. e.g. [Sto]).

In Corollary 1.17, we may consider in particular the trivial metric on each factor so that the ℓ^1-metric on $X = X_1 \times \cdots \times X_n$ is the *Hamming metric*

$$d(x, y) = \sum_{i=1}^n \mathbf{1}_{\{x_i \neq y_i\}} = \operatorname{Card}\{1 \leq i \leq n; x_i \neq y_i\}$$

(where we denote by $x = (x_1, \ldots, x_n)$ and $y = (y_1, \ldots, y_n)$ the coordinates of the points x, y in the product space X). Hence, any product probability measure P on the product space has normal concentration

$$\alpha_P(r) \leq \mathrm{e}^{-r^2/8n}, \quad r > 0. \tag{1.24}$$

This is already good enough to recover (up to numerical constants) concentration (1.5) on the discrete cube $\{0, 1\}^n$. In terms of the observable diameter of the product space $X = X_1 \times \cdots \times X_n$ with respect to the Hamming metric and any product probability measure P,

$$\operatorname{ObsDiam}_P(X) = O(\sqrt{n}). \tag{1.25}$$

As already mentioned next to Proposition 1.11, this approach is however not well suited to concentration bounds with respect to ℓ^2-metrics (like the Euclidean metric) which are independent of the number of spaces in the product (dimension free concentration) as it is the case for Gaussian measures.

In Chapter 4, we establish normal concentration by showing that for every Borel set A in X,

$$\int \mathrm{e}^{cd(\cdot, A)^2} d\mu \leq \frac{1}{\mu(A)} \tag{1.26}$$

where $d(x, A)$ is the distance from the point x to the set A. Indeed, under (1.26), for every $r > 0$,

$$1 - \mu(A_r) = \mu(\{d(\cdot, A) \geq r\}) \leq \frac{1}{\mu(A)} \mathrm{e}^{-cr^2}$$

so that $\alpha_\mu(r) \leq 2\,\mathrm{e}^{-cr^2}$, $r > 0$.

In the same spirit, one may investigate infimum-convolution inequalities that actually allow us to investigate concentration properties outside the usual metric setting. Given a measurable space $(X, \mathcal{A})$, consider a non-negative *cost function* $\tilde{c} : X \times X \to \mathbb{R}_+$. A typical example is the quadratic cost $\tilde{c}(x, y) = \tilde{c}(x - y) = \frac{1}{2}|x - y|^2$ on $\mathbb{R}^n \times \mathbb{R}^n$. Given a real-valued measurable function f on X, denote by $Q_{\tilde{c}} f$ the *infimum-convolution* of f with respect to the cost $\tilde{c}$ defined by

$$Q_{\tilde{c}} f(x) = \inf_{y \in X} \big[f(y) + \tilde{c}(x, y)\big], \quad x \in X. \tag{1.27}$$

If μ is a probability measure on $\mathcal{A}$, and $\tilde{c}$ a non-negative measurable cost function on X, we say that μ satisfies an *infimum-convolution inequality* with respect to the cost $\tilde{c}$ if for all bounded measurable functions f on X,

$$\int e^{Q_{\tilde{c}} f} d\mu \int e^{-f} d\mu \leq 1. \tag{1.28}$$

If we adopt the convention that $+\infty \times 0 \leq 1$, the above inequality extends to all $\overline{\mathbb{R}}$-valued functions f.

The preceding definition is motivated by its connection to concentration.

Proposition 1.18. *If μ satisfies an infimum-convolution inequality with respect to the cost $\tilde{c}$, then, for every measurable set A in X and every $r > 0$,*

$$1 - \mu\big(\{\inf_{y \in A} \tilde{c}(\cdot, y) < r\}\big) \leq \frac{1}{\mu(A)} e^{-r}.$$

The proof simply follows by applying (1.28) to the function f that is equal to 0 on A and $+\infty$ outside. Then $Q_{\tilde{c}} f(x) \geq r$ if and only if $\inf_{y \in A} \tilde{c}(x, y) \geq r$. Conclude with Chebyshev's exponential inequality.

A typical application is of course the cost given by the distance $\tilde{c}(x, y) = d(x, y)$ on a metric space (X, d). One may also consider powers of the distance functions. The example of

$$\tilde{c}(x, y) = \frac{c}{2}\, d(x, y)^2, \quad x, y \in X, \tag{1.29}$$

for some $c > 0$ is of particular interest. It should be noted that by Jensen's inequality, inequality (1.28) implies that for every bounded measurable function f on X,

$$\int e^{Q_{\tilde{c}} f} d\mu \leq e^{\int f d\mu}. \tag{1.30}$$

Now, for the choice (1.29) of the cost $\tilde{c}$, whenever F is Lipschitz on (X, d) with Lipschitz coefficient $\|F\|_{\mathrm{Lip}}$, for every $x \in X$,

$$\begin{aligned} Q_{\tilde{c}} F(x) &\geq F(x) + \inf_{y \in X} \Big[-\|F\|_{\mathrm{Lip}}\, d(x, y) + \frac{c}{2}\, d(x, y)^2 \Big] \\ &\geq F(x) - \frac{1}{2c}\, \|F\|_{\mathrm{Lip}}^2. \end{aligned}$$

Hence it follows from (1.30) that

$$\int e^F d\mu \leq e^{\int F d\mu + \|F\|_{\mathrm{Lip}}^2 / 2c}.$$

We thus recover from Proposition 1.18 the concentration result deduced in Proposition 1.14 from the Laplace bounds. We will return to this aspect in Chapter 6.

Another instance of interest is the case of a topological vector space X with its Borel σ-field. Given a measurable non-negative function $\tilde{c}$ on X, consider the cost $\tilde{c}(x, y) = \tilde{c}(x - y)$, $x, y \in X$. The concentration result of Proposition 1.18 then amounts to the inequality

$$1 - \mu\big(\{A + \{\tilde{c} < r\}\}\big) \leq \frac{1}{\mu(A)} e^{-r}, \quad r > 0 \tag{1.31}$$

where we denote by $A + B$ the Minkowski sum

$$A + B = \{x + y; x \in A, y \in B\}$$

of two sets A and B in X.

As for the concentration functions and Laplace functionals, the infimum-convolution property (1.28) satisfies a useful contraction property. Indeed, let μ satisfy an infimum-convolution inequality with respect to a cost function $\tilde{c}$ on $X \times X$. Let $\tilde{e}$ be a cost function on $Y \times Y$ and let $\varphi : X \to Y$ be such that

$$\tilde{e}\big(\varphi(x), \varphi(y)\big) \leq \tilde{c}(x, y)$$

for all $x \in X$, $y \in Y$. Then the image measure of μ by φ satisfies an infimum-convolution inequality with cost $\tilde{e}$ on Y. Indeed, it is enough to observe that for every $f : Y \to \mathbb{R}$, $Q_{\tilde{c}}(f \circ \varphi) \geq (Q_{\tilde{e}} f) \circ \varphi$.

One basic aspect of the infimum-convolution inequality is its stability under products.

Proposition 1.19. *Let μ and ν be Borel probability measures satisfying infimum-convolution inequalities with respect to costs $\tilde{c}$ and $\tilde{e}$ on X and Y, respectively. Then $\mu \otimes \nu$ satisfies an infimum-convolution inequality with respect to the cost $\tilde{c} + \tilde{e}$ on $X \times Y$.*

Proof. Given $f = f(x, y)$ on the product space, set $f^y(x) = f(x, y)$ for every $y \in Y$ as well as $g(y) = \log \int e^{Q_{\tilde{c}} f^y} d\mu$. Observe that

$$\int e^{Q_{\tilde{c}+\tilde{e}} f} d\mu \leq e^{Q_{\tilde{e}} g}.$$

By (1.28) applied to f^y with respect to μ for every y and to g with respect to ν, we have

$$\int e^{-f^y} d\mu \leq e^{-g(y)} \quad \text{and} \quad \int e^{Q_{\tilde{e}} g} d\nu \int e^{-g} d\nu \leq 1.$$

The conclusion follows. □

As a consequence of Proposition 1.19, each time a measure μ satisfies the infimum-convolution inequality (1.28), the n-fold product probability measure μ^n on X^n satisfies the concentration inequality of Proposition 1.18 with respect to the sum of the costs along each coordinate. With respect to Proposition 1.15, this result allows us to deal with more general metrics than the ℓ^1-metric on product spaces, such as for example the ℓ^2-metric related to the quadratic cost. In particular, if $P = \mu_1 \times \cdots \times \mu_n$ is any product measure on $\mathbb{R}^n$, in order that P has normal concentration with respect to the Euclidean metric, it suffices to know that each one dimensional marginal μ_i satisfies an infimum-convolution inequality (1.28) with the same cost $\tilde{c}(x) = cx^2$, $x \in \mathbb{R}$, for some $c > 0$. We develop application of this principle to dimension free concentration in Chapters 4 and 6.

Notes and Remarks

The definition of concentration function was first introduced in [Am-M]. It is formalized in [Gr-M1] and further analyzed in [M-S], and these are the references from which the elementary properties of Section 1.2 are taken. Concentration and deviation inequalities for Lipschitz functions on high dimenisonal spheres have been emphasized by P. Lévy [Lé]. Their equivalence with concentration inequalities is also part of the folklore of the subject. Sharper constants in normal deviation inequalities are discussed in [Bob6]. Many properties discussed in Section 1.3 of this chapter may be found in the reference [M-S]. Further introductions to the concentration of measure phenomenon and its applications are [Mi5] and [Sche5]. See also [Bal] for an introduction in the context of modern convex geometry.

Observable diameters of metric measure spaces are described by M. Gromov in [Grom2] where further geometric invariants related to concentration are analyzed. Expansion coefficients are also discussed there and used in [Gr-M1] (see Chapter 3).

Inequalities on Laplace transforms are a traditional tool to produce (normal) concentration. Infimum-convolution inequalities were introduced in this way by B. Maurey in [Mau2] inspired by investigations by M. Talagrand [Tal3]. The various results on concentration under infimum-convolution inequalities of Section 1.6 are taken from [Mau2] and will be revisited in Chapter 6.

2. ISOPERIMETRIC AND FUNCTIONAL EXAMPLES

As already illustrated in the first section of the first chapter, isoperimetric inequalities are a basic source of examples of concentration properties. In the first sections of this chapter we present the basic isoperimetric and Brunn-Minkowski inequalities leading to measure concentration (and in particular, cover more carefully the early examples of Section 1.1). We however do not present proofs of the isoperimetric inequalities but rather provide in Sections 2.2 and 2.3 complete functional arguments (mainly from semigroup theory) to reach the concentration examples initially derived from isoperimetry. Indeed, while the first concern of isoperimetry is extremal sets, the concentration phenomenon dealing with non-infinitesimal neighborhoods is a milder property that may be established far outside the isoperimetric context as will be amply demonstrated throughout this book. Geometric examples involve lower bounds on the Ricci curvature as parts of the Riemannian comparison theorems. These are treated here with rather elementary functional tools (Bochner's formula and convexity criteria) which do not require any deep knowledge in Riemannian geometry.

2.1 Isoperimetric examples

Isoperimetric inequalities are a basic source of examples of the concentration principle. However, their main interest lies in extremal sets and surface measures, whereas concentration deals with big enlargements. As such, concentration covers situations far outside isoperimetric considerations. This section thus only describes the concentration properties that follow from isoperimetry rather than the isoperimetric inequalities themselves.

Let (X, d) be a metric space equipped with a (not necessarily finite) Borel measure μ. The *boundary measure* or Minkowski content of a Borel set A in X with respect to μ is defined as

$$\mu^+(A) = \liminf_{r \to 0} \frac{1}{r}\, \mu(A_r \backslash A) \tag{2.1}$$

where we recall that $A_r = \{x \in X; d(x, A) < r\}$ is the (open) r-neighborhood of A (with respect to d).

The *isoperimetric function* of μ is the largest function I_μ on $[0, \mu(X)]$ such that

$$\mu^+(A) \geq I_\mu\big(\mu(A)\big) \tag{2.2}$$

holds for every Borel set $A \subset X$ such that $\mu(A) < \infty$. When B is such that $\mu^+(B) = I_\mu(\mu(B))$, B has minimal boundary measure among sets of the same measure, and B is said to be an *extremal set*. Isoperimetry thus expresses that whenever A is a measurable set with the same volume as an extremal set B, then the boundary measure of A is greater than or equal to the boundary measure of B. One of the main interests in isoperimetric inequalities lies in explicit expressions for extremal sets. Unfortunately, extremal sets are rather difficult to determine. One motivation for the concentration phenomenon is that methods and results are available for large families of examples for which the characterization of isoperimetric extremal sets is simply hopeless.

The isoperimetric function I_μ is explicitly known only in a few cases. The most notable ones are the constant curvature spaces, going back to the works of P. Lévy [Lé] and E. Schmidt [Schmi]. Indeed, let X be either the Euclidean n-space $\mathbb{R}^n$, the standard n-sphere $\mathbb{S}^n \subset \mathbb{R}^{n+1}$ with its geodesic metric or the n-dimensional hyperbolic space $\mathbb{H}^n$ with its hyperbolic metric. Equip these spaces with their Riemannian volume element $dv = d\mu$ (normalized in the case of the sphere) and denote by $v(r)$ the measure of a ball with radius $r > 0$. Then

$$I_\mu = v' \circ v^{-1}. \tag{2.3}$$

In particular, $I_\mu(a) = n\omega_n^{1/n} a^{(n-1)/n}$ in the case of Lebesgue measure on $\mathbb{R}^n$ where ω_n is the volume of the Euclidean unit ball. Extremal sets are given in each case by geodesic balls. Further examples will be discussed later. One of the most fruitful among such examples is Gaussian isoperimetry, which will be described below as limiting spherical isoperimetry.

The first statement is the bridge between isoperimetric and concentration properties. We assume there for simplicity that μ is Borel measure on a metric space (X, d) such that the lim inf in the definition (2.1) of $\mu^+(A)$ is a true limit for A given by a finite union of open balls and such that the family of these subsets is a determining class for μ. These assumptions are easily seen to cover the preceding classical examples, as well as measures which are absolutely continuous with respect to Lebesgue measure on $\mathbb{R}^n$.

Proposition 2.1. *Let μ be as above. Assume that $I_\mu \geq v' \circ v^{-1}$ for some strictly increasing differentiable function $v : I \subset \mathbb{R} \to [0, \mu(X)]$. Then, for every $r > 0$,*

$$v^{-1}\big(\mu(A_r)\big) \geq v^{-1}\big(\mu(A)\big) + r.$$

Proof. By the hypotheses, it is enough to assume that A is given by a finite union of open balls. The family of such sets is closed under the operation $A \mapsto A_r$, $r > 0$. Now, the function $h(r) = v^{-1}(\mu(A_r))$ satisfies

$$h'(r) = \frac{\mu^+(A_r)}{v' \circ v^{-1}(\mu(A_r))} \geq 1$$

so that $h(r) = h(0) + \int_0^r h'(s)ds \geq v^{-1}(\mu(A)) + r$ which is the desired claim. □

Note that conversely, if

$$v^{-1}\big(\mu(A_r)\big) \geq v^{-1}\big(\mu(A)\big) + r, \quad r > 0,$$

then, for every subset A with $\mu(A) < \infty$,

$$\begin{aligned} \mu^+(A) &\geq \liminf_{r\to 0} \frac{1}{r} \Big[v\Big(v^{-1}\big(\mu(A)\big) + r\Big) - \mu(A)\Big] \\ &= v' \circ v^{-1}\big(\mu(A)\big). \end{aligned} \tag{2.4}$$

For the constant curvature spaces, equality in (2.4) is achieved on geodesic balls so that Proposition 2.1 may be expressed equivalently as the comparison result

$$\mu(A_r) \geq \mu(B_r), \quad r > 0, \tag{2.5}$$

as soon as A is a set with $\mu(A) = \mu(B)$ where B is a ball. This geometric description emphasizes the interest in explicit knowledge of extremal sets to bound below $\mu(A_r)$ for any set A. However, the difference in perspective between isoperimetry and concentration is that the latter makes use of the integrated form (2.5) of isoperimetry with no emphasis on surface measure and extremal sets.

If μ is a probability measure on the Borel sets of (X, d), recall its concentration function $\alpha_\mu = \alpha_{(X,d,\mu)}$.

Corollary 2.2. *Let μ be a probability measure on (X, d) for which Proposition 2.1 applies and assume that $I_\mu \geq v' \circ v^{-1}$. Then,*

$$\alpha_{(X,d,\mu)}(r) \leq 1 - v\big(v^{-1}(\tfrac{1}{2}) + r\big), \quad r > 0.$$

Denote for example by σ^n the normalized volume element on the unit sphere $\mathbb{S}^n$ equipped with its geodesic distance, $n \geq 2$. Then, for every $0 < r \leq \pi$,

$$v(r) = s_n^{-1} \int_0^r \sin^{n-1} \theta d\theta$$

where $s_n = \int_0^\pi \sin^{n-1} \theta d\theta$. Since $v^{-1}(\frac{1}{2}) = \frac{\pi}{2}$, we evaluate the quantity $1 - v(\frac{\pi}{2} + r)$ for $0 < r \leq \frac{\pi}{2}$. We have

$$\begin{aligned} 1 - s_n^{-1} \int_0^{(\pi/2)+r} \sin^{n-1}\theta d\theta &= s_n^{-1} \int_{(\pi/2)+r}^{\pi} \sin^{n-1}\theta d\theta \\ &= s_n^{-1} \int_r^{\pi/2} \cos^{n-1}\theta d\theta. \end{aligned}$$

By the change of variables $\theta = \tau/\sqrt{n-1}$, and the elementary inequality $\cos u \leq \mathrm{e}^{-u^2/2}$, $0 \leq u \leq \frac{\pi}{2}$,

$$\begin{aligned} \int_r^{\pi/2} \cos^{n-1}\theta d\theta &= \frac{1}{\sqrt{n-1}} \int_{r\sqrt{n-1}}^{(\pi/2)\sqrt{n-1}} \cos^{n-1}\Big(\frac{\tau}{\sqrt{n-1}}\Big) d\tau \\ &\leq \frac{1}{\sqrt{n-1}} \int_{r\sqrt{n-1}}^{\infty} \mathrm{e}^{-\tau^2/2} d\tau \\ &\leq \frac{\sqrt{\pi}}{\sqrt{2(n-1)}}\, \mathrm{e}^{-(n-1)r^2/2} \end{aligned}$$

where we used that $1 - \Phi(r) \leq \frac{1}{2}\,\mathrm{e}^{-r^2/2}$, $r \geq 0$. To evaluate s_n from below, note that by integration by parts $s_n = [(n-2)/(n-1)]s_{n-2}$, $n \geq 2$, so that $\sqrt{n-1}\,s_n \geq \sqrt{n-3}\,s_{n-2}$. Hence $\sqrt{n-1}\,s_n \geq 2$, $n \geq 2$. As a consequence, we may state the following important consequence.

Theorem 2.3. *For the standard n-sphere* $\mathbb{S}^n$, $n \geq 2$, *equipped with its geodesic metric d and normalized volume element* μ,

$$\alpha_{(\mathbb{S}^n, d, \sigma^n)} \leq \mathrm{e}^{-(n-1)r^2/2}, \quad r > 0$$

(actually $0 < r \leq \pi$*).*

Up to numerical constants, we may regard $\mathbb{S}^n$ equivalently as a subset of $\mathbb{R}^{n+1}$ endowed with the usual Euclidean metric. Notice that Theorem 2.3 applied to sets the diameter of which tends to zero implies the classical isoperimetric inequality in Euclidean space.

As already emphasized in Section 1.1, this property is of special interest when the dimension n is large in which case a very small enlargement, of the order of $\frac{1}{\sqrt{n}}$, yields a set of almost full measure. In other words, given a set A with $\sigma^n(A) \geq \frac{1}{2}$, most of the points on the sphere are within distance $\frac{1}{\sqrt{n}}$ from A. By Proposition 1.3, for every continuous function F on $\mathbb{S}^n$ and every $\eta > 0$,

$$\mu\big(\{|F - m_F| > \omega_F(\eta)\}\big) \leq 2\,\mathrm{e}^{-(n-1)\eta^2/2} \tag{2.6}$$

where m_F is a median of F and ω_F the modulus of continuity of F. When n is large, functions with small oscillations are thus almost constant. Inequality (2.6) is referred to as Lévy's inequality. In terms of the observable diameter (cf. Section 1.4),

$$\mathrm{ObsDiam}_{\sigma^n}(\mathbb{S}^n) = O\Big(\frac{1}{\sqrt{n}}\Big). \tag{2.7}$$

For the n-sphere $\mathbb{S}^n_R$ with radius $R > 0$ equipped with the normalized invariant measure σ^n_R,

$$\alpha_{\sigma^n_R}(r) \leq \mathrm{e}^{-(n-1)r^2/2R^2}, \quad r > 0. \tag{2.8}$$

The isoperimetric inequality on spheres has been extended by M. Gromov [Grom1], using ideas going back to P. Lévy, as a comparison theorem for Riemannian manifolds with strictly positive curvature. Let (X, g) be a compact connected smooth Riemannian manifold of dimension n (≥ 2) with Riemannian metric g, equipped with the normalized Riemannian volume element $d\mu = \frac{dv}{V}$ where V is the total volume of X. Denote by $c(X)$ the infimum of the Ricci curvature tensor over all unit tangent vectors, and assume that $c(X) > 0$. The n-sphere $\mathbb{S}^n_R$ with radius $R > 0$ is of constant curvature $c(\mathbb{S}^n_R) = \frac{n-1}{R^2}$. Ricci curvature is a way to describe the variations of the Riemannian measure with respect to the Euclidean one. We refer to [C-E], [G-H-L], [Cha2], etc. for standard introductions to curvature in Riemannian geometry as well as for some classical examples (see also below). While Ricci curvature appears as a crucial geometric parameter in the subsequent statements, we provide in Section 2.3 a simple functional approach to these results relying only on the use of Bochner's formula (2.29) that does not require any deep understanding of Riemannian geometry. Furthermore, the log-concavity condition of Theorem 2.7 and Proposition 2.18 below may be used to get some further insight into the geometric content of curvature in this context. Recall the isoperimetric function I.

Theorem 2.4. *Let (X, g) be a compact connected smooth Riemannian manifold of dimension n (≥ 2) equipped with the normalized Riemannian volume element $d\mu = \frac{dv}{V}$ such that $c(X) > 0$. Then*

$$I_\mu \geq I_{\sigma_R^n}$$

where $R > 0$ is such that $c(\mathbb{S}_R^n) = \frac{n-1}{R^2} = c(X)$. In particular, (X, g, μ) has normal concentration

$$\alpha_{(X,g,\mu)}(r) \leq \mathrm{e}^{-cr^2/2}, \quad r > 0.$$

This isoperimetric comparison theorem strongly emphasizes the importance of a model space, here the sphere, to which manifolds may be compared. It does not yield any information on the extremal sets for I_μ itself. Equality in the inequality $I_\mu \geq I_{\sigma_R^n}$ occurs only if X is a sphere.

Theorem 2.4 includes a number of geometric examples of interest for which a lower bound on the Ricci curvature is known, some of which we discuss now.

It is clear that (Riemannian) products of spheres or of Riemannian manifolds with a common lower bound $c > 0$ on the Ricci curvatures satisfy Theorem 2.4. Therefore measure concentration holds independently of the number of factors in the product. This is in contrast with ℓ^1-products as analyzed in Proposition 1.11 and Corollary 1.17.

Let $\mathbb{O}^n$ be the group of all $n \times n$ real orthogonal matrices. For $1 \leq k \leq n$, let

$$\mathbb{W}_k^n = \{e = (e_1, \ldots, e_k); e_i \in \mathbb{R}^n, e_i \cdot e_j = \delta_{ij}, 1 \leq i, j \leq k\}$$

be the so-called Stiefel manifolds. Equip $\mathbb{W}_k^n$ with the metric

$$d(e, f) = \Big(\sum_{i=1}^k |e_i - f_i|^2\Big)^{1/2}.$$

Note that $\mathbb{W}_n^n = \mathbb{O}^n$, $\mathbb{W}_1^n = \mathbb{S}^{n-1}$ and $\mathbb{W}_{n-1}^n = \mathbb{SO}^n = \{T \in \mathbb{O}^n; \det(T) = 1\}$. In general, $\mathbb{W}_{n,k}$ may be identified with the quotient $\mathbb{O}^n/\mathbb{O}^{n-k}$ via the map $\varphi : \mathbb{O}^n \to \mathbb{W}_k^n$, $\varphi(e_1, \ldots, e_n) = (e_1, \ldots, e_k)$.

The value of $c(\mathbb{SO}^n)$ is known and equal to $\frac{n-1}{4}$ [C-E]. Therefore, by Theorem 2.4, the normalized Haar measure μ on $\mathbb{SO}^n$ has normal concentration function $\mathrm{e}^{-(n-1)r^2/8}$. By Proposition 1.2, this property is inherited by the quotients of $\mathbb{SO}^n$. In particular, all the Stiefel manifolds have normal concentration. Up to normalization factors, the same conclusions apply to the unitary group.

Similar conclusions hold for Grassmann manifolds. Recall that the Grassmann manifold $\mathbb{G}_k^n$, $1 \leq k \leq n$, is a metric space of all k-dimensional subspaces of $\mathbb{R}^n$ with the Hausdorff distance between the unit spheres of the subspaces of E and F:

$$d(E, F) = \sup\{d(x, \mathbb{S}^{n-1} \cap E); x \in \mathbb{S}^{n-1} \cap F\}.$$

Since $\mathbb{G}_k^n$ is again a quotient of $\mathbb{SO}^n$, (normalized Haar) measure on $\mathbb{G}_k^n$ has normal concentration.

A wealth of further geometric examples are presented and discussed in [Grom2] on the basis of Theorem 2.4, including subvarieties in $\mathbb{S}^n$, real and complex projectives spaces, etc. For example, the (Lipschitz) Hopf fibration $\mathbb{S}^{2n+1} \to \mathbb{C}\mathrm{P}^n$ indicates that

$$\mathrm{ObsDiam}(\mathbb{C}\mathrm{P}^n) = O\Big(\frac{1}{\sqrt{n}}\Big).$$

Of more delicate algebraic and metric geometry, the observable diameter of a complex algebraic submanifold $X \subset \mathbb{C}\mathrm{P}^n$ of fixed codimension and degree, with normalized Riemannian volume and the induced path metric, satisfies

$$\mathrm{ObsDiam}(X) = O\Big(\frac{\log n}{n}\Big).$$

We refer to [Grom2] for further details.

It is well known that uniform measures on n-dimensional spheres with radius $\sqrt{n}$ approximate (when projected on a finite number of coordinates) Gaussian measures (Poincaré's lemma) [MK], [Str2], [Le3]. In this sense, the isoperimetric inequality on high dimensional spheres gives rise to an isoperimetric inequality for Gaussian measures. Extremal sets are then half-spaces (which may be considered as balls with centers at infinity). More precisely, let $\gamma = \gamma^k$ be the canonical Gaussian measure on the Borel sets of $\mathbb{R}^k$ with density $(2\pi)^{-k/2}\mathrm{e}^{-|x|^2/2}$ with respect to Lebesgue measure. Equip $\mathbb{R}^k$ with its usual Euclidean metric induced by the norm $|x| = \big(\sum_{i=1}^k x_i^2\big)^{1/2}$, $x = (x_1, \ldots, x_k) \in \mathbb{R}^k$.

Theorem 2.5. *For any $k \geq 1$,*

$$I_\gamma = \Phi' \circ \Phi^{-1}$$

where $\Phi(t) = (2\pi)^{-1/2}\int_{-\infty}^t \mathrm{e}^{-x^2/2}dx$, $t \in [-\infty, +\infty]$, is the distribution function of the canonical Gaussian measure in dimension one. Moreover, the equality $\gamma^+(A) = I_\gamma(\gamma(A))$ holds if and only if A is a half-space in $\mathbb{R}^k$.

Since the distribution of half-spaces is one-dimensional, the isoperimetric inequality of Theorem 2.5 in its integrated form (2.5) indicates that whenever A is a Borel set in $\mathbb{R}^k$ and $\gamma(A) = \Phi(a)$ for some $a \in \mathbb{R}$, then

$$\gamma(A_r) \geq \Phi(a + r) \tag{2.9}$$

for every $r > 0$ where A_r is the neighborhood of order $r > 0$ in the Euclidean metric on $\mathbb{R}^k$. Using that $\Phi(0) = \frac{1}{2}$ and $1 - \Phi(r) \leq \mathrm{e}^{-r^2/2}$, $r > 0$, we get the following corollary.

Corollary 2.6. *The canonical Gaussian measure γ on $\mathbb{R}^k$ equipped with its Euclidean metric has normal concentration*

$$\alpha_{(\mathbb{R}^k,\gamma)}(r) \leq \mathrm{e}^{-r^2/2}, \quad r > 0.$$

Alternatively, one may perform the Poincaré limit on (2.8). In particular, by Proposition 1.3, for every 1-Lipschitz function F on $\mathbb{R}^k$ and every $r \geq 0$,

$$\gamma(\{F \geq m_F + r\}) \leq \mathrm{e}^{-r^2/2} \tag{2.10}$$

where m_F is a median of F for γ. This inequality extends the trivial case of linear functions. In terms of the observable diameter,

$$\mathrm{ObsDiam}_\gamma(\mathbb{R}^k) = O(1) \tag{2.11}$$

uniformly in $k \geq 1$. The preceding results for canonical Gaussian measures immediately extend to arbitrary Gaussian measures replacing the standard Euclidean structure by the corresponding covariance structure. One important feature of both Theorem 2.5 and Corollary 2.6 is indeed the dimension free character of the isoperimetric and concentration functions of Gaussian measures. This observation opens the extension to infinite dimensional Gaussian analysis (cf. [Bor2], [Le-T], [Le3], [Li], [Bog]). For example, (2.10) holds similarly for a functional $F : E \to \mathbb{R}$ on an abstract Wiener space (E, H, μ) that is 1-Lipschitz in the directions of H, that is, $|F(x+h) - F(x)| \leq |h|$ for all $x \in E$, $h \in H$.

As for Theorem 2.4, Theorem 2.5 may be turned into a comparison theorem. Due to the dimension free character of Gaussian isoperimetry, this is better presented in terms of (strictly) log-concave probability measures (on a finite dimensional state space). The following statement is moreover a concrete illustration of the geometric content of Ricci curvature in the preceding Riemannian language.

Theorem 2.7. *Let μ be a probability measure on $\mathbb{R}^n$ with smooth strictly positive density e^{-U} with respect to Lebesgue measure such that, for some $c > 0$, $\mathrm{Hess}\, U(x) \geq c\,\mathrm{Id}$ as symmetric matrices uniformly in $x \in \mathbb{R}^n$. Then,*

$$I_\mu \geq \sqrt{c}\, I_\gamma\,.$$

In particular,

$$\alpha_{(\mathbb{R}^n,\mu)}(r) \leq \mathrm{e}^{-cr^2/2}, \quad r > 0,$$

where $\mathbb{R}^n$ is equipped with the standard Euclidean metric, as usual.

The convexity condition in Theorem 2.7 is stable under various operations. For example, image measures of log-concave measures in convex domains in $\mathbb{R}^n$ under linear projections $\mathbb{R}^n \to \mathbb{R}^m$ are log-concave with the same log-concavity bound.

Although Theorem 2.3 may be properly modified to yield similarly the case of dimension one, that is, the case of the circle, it is simpler to deduce this case by contraction of the Gaussian measure following Proposition 1.2. We may even state the result on the product space. Let $\mu_{[0,1]^n}$ be the uniform measure on $[0,1]^n$ (with the induced Euclidean metric) that is the product measure of Lebesgue measure on $[0,1]$ in each direction. Then $\mu_{[0,1]^n}$ is the image of the canonical Gaussian measure $\gamma = \gamma^n$ on $\mathbb{R}^n$ under the map $\varphi = \Phi^{\otimes n}$ where we recall that Φ is the distribution function of the standard normal distribution in dimension one. Since $\|\varphi\|_{\mathrm{Lip}} = (2\pi)^{-1/2}$, we may state the following consequence. The constants are not sharp.

Proposition 2.8. *Let $\mu_{[0,1]^n}$ be a uniform measure on $[0,1]^n$. Then*

$$\alpha_{([0,1]^n,\mu_{[0,1]^n})}(r) \leq \mathrm{e}^{-\pi r^2}, \quad r > 0.$$

We may also push forward Gaussian measure on the Euclidean unit ball.

Proposition 2.9. *Let $\mathcal{B}^n$ be the Euclidean unit ball in $\mathbb{R}^n$ equipped with the normalized uniform measure $\mu_{\mathcal{B}^n}$. Then*

$$\alpha_{(\mathcal{B}^n,\mu_{\mathcal{B}^n})}(r) \le \mathrm{e}^{-cnr^2}, \quad r > 0,$$

where $c > 0$ is numerical.

Proof. Denote by γ_ρ the image measure of the canonical Gaussian measure γ on $\mathbb{R}^n$ under the 1-Lipschitz map $x \mapsto |x|$. Further, denote by $v : \mathbb{R}^+ \to [0,1]$ the map that pushes γ_ρ forward to the probability measure $d(t^n)$ on $[0,1]$. In other words, for every $t \in [0,1]$,

$$\gamma\Big(\big\{x \in \mathbb{R}^n; v\big(|x|\big) \le t\big\}\Big) = \gamma_\rho\big(\{r \ge 0; v(r) \le t\}\big) = t^n.$$

It is an easy matter to check from the definition that $\|v\|_{\mathrm{Lip}} \le \frac{C}{\sqrt{n}}$ for some numerical $C > 0$. Then set

$$\varphi(x) = \frac{x}{|x|}\, v\big(|x|\big), \quad x \in \mathbb{R}^n \setminus \{0\}.$$

The image measure of γ under φ is $\mu_{\mathcal{B}^n}$ and $\|\varphi\|_{\mathrm{Lip}} = \|v\|_{\mathrm{Lip}}$. The conclusion then follows again from Proposition 1.2. □

One may also deduce Proposition 2.9 from concentration on spheres by integration in polar coordinates together with the fact (see above) that the measure $d(t^n)$ on $[0,1]$ has normal concentration with rate of the order of n. Concentration on Euclidean spheres or balls is thus similar, with the same dimensional rate. Actually, simple volume estimates show that a uniform measure on the Euclidean unit ball $\mathcal{B}^n$ in high dimension is strongly concentrated on its boundary since for $\varepsilon \sim \frac{t}{n}$,

$$\mathrm{vol}_n(\mathcal{B}^n) - (1-\varepsilon)^n \mathrm{vol}_n(\mathcal{B}^n) \sim (1 - \mathrm{e}^{-t})\, \mathrm{vol}_n(\mathcal{B}^n)$$

as $n \to \infty$. Here $\mathrm{vol}_n(\cdot)$ denotes the volume element in $\mathbb{R}^n$ (Lebesgue measure). The same holds for large families of convex bodies, including the cube. This observation intuitively justifies many transfers from the ball to the sphere and conversely.

In the same spirit, one may transfer concentration for Gaussian measures back to spheres. This is the content of the following proposition, which we state quite informally.

Proposition 2.10. *Concentration for Gaussian measures implies concentration on spheres.*

Proof. The image measure of the canonical Gaussian measure on $\mathbb{R}^{n+1}$ under the map $x \mapsto \frac{x}{|x|}$ is σ^n. Given a 1-Lipschitz function $F : \mathbb{S}^n \to \mathbb{R}$, and x_0 a fixed point on σ^n, define $G : \mathbb{R}^{n+1} \to \mathbb{R}$ by $G(x) = |x|\big(F\big(\frac{x}{|x|}\big) - F(x_0)\big)$ $(G(0) = 0)$. Then $G = F - F(x_0)$ on $\mathbb{S}^n$ and $\|G\|_{\mathrm{Lip}} \le 2\pi$. Let m_n be a median of $x \mapsto |x|$ with respect to γ on $\mathbb{R}^{n+1}$. For every $r > 0$,

$$\begin{aligned}
\gamma \otimes \gamma\Big(\Big\{(x,y) \in \mathbb{R}^{n+1} \times \mathbb{R}^{n+1}; \Big|G\Big(\frac{x}{|x|}\Big) - G\Big(\frac{y}{|y|}\Big)\Big| \ge 3r\Big\}\Big)& \\
\le \gamma \otimes \gamma\Big(\Big\{(x,y); \Big|G\Big(\frac{x}{m_n}\Big) - G\Big(\frac{y}{m_n}\Big)\Big| \ge r\Big\}\Big)& \\
+ 2\gamma\Big(\Big\{x; \Big|G\Big(\frac{x}{|x|}\Big) - G\Big(\frac{x}{m_n}\Big)\Big| \ge r\Big\}\Big).&
\end{aligned}$$

Now,

$$\left|G\Big(\frac{x}{|x|}\Big) - G\Big(\frac{x}{m_n}\Big)\right| \le 2\pi \left|\frac{|x|}{m_n} - 1\right|$$

and $G(\frac{\cdot}{m_n})$ is $\frac{2\pi}{m_n}$-Lipschitz on $\mathbb{R}^{n+1}$. Apply now (2.10) to both $G(\frac{\cdot}{m_n})$ and $x \mapsto \frac{|x|}{m_n}$ to get

$$\gamma \otimes \gamma\bigg(\bigg\{(x,y);\left|G\Big(\frac{x}{|x|}\Big) - G\Big(\frac{y}{|y|}\Big)\right| \ge 3r\bigg\}\bigg)$$
$$\le 2\,\mathrm{e}^{-m_n^2 r^2/32\pi^2} + 4\,\mathrm{e}^{-m_n^2 r^2/8\pi^2}.$$

A simple computation shows that $m_n \sim \sqrt{n}$. The claim easily follows by Corollary 1.5. □

We close this section with some discrete examples. We start by a further isoperimetric inequality, on the discrete cube, and with the corresponding concentration phenomenon. Equip the n-dimensional discrete cube $X = \{0,1\}^n$ with the normalized Hamming metric

$$d(x,y) = \frac{1}{n}\,\mathrm{Card}\big(\{x_i \ne y_i; i = 1,\dots,n\}\big),$$

$x = (x_1,\dots,x_n), y = (y_1,\dots,y_n) \in \{0,1\}^n$, and with the uniform measure $\mu(A) = 2^{-n}\,\mathrm{Card}\,(A)$, $A \subset X$. Extremal sets of the isoperimetric problem

$$\inf\{\mu(A_r); \mu(A) = \varepsilon\}$$

have been identified in [Ha] as *hereditary* sets A in the sense that if $x = (x_1,\dots,x_n) \in A$ and if $y = (y_1,\dots,y_n) \in \{0,1\}^n$ is such that $y_i \le x_i$ for all $i = 1,\dots,n$, then $y \in A$. For $\varepsilon = \frac{1}{2}$, such an hereditary set A is given by

$$A = \bigg\{x \in \{0,1\}^n; \sum_{i=1}^n x_i \le \frac{n}{2}\bigg\}.$$

For any integer $k \ge 0$, classical binomial bounds (cf. e.g. [MD2]) show that

$$\mu\bigg(\bigg\{x; \sum_{i=1}^n x_i - \frac{n}{2} \ge k\bigg\}\bigg) \le \mathrm{e}^{-2k^2/n}.$$

We may thus come to the following concentration result on the discrete cube. With a weaker numerical constant, this was already observed through (1.24). The normalization of the metric so far is chosen for the matter of comparison with spherical concentration.

Theorem 2.11. *For a uniform measure $\mu = \mu^n$ on $\{0,1\}^n$ equipped with the normalized Hamming metric,*

$$\alpha_{(\{0,1\}^n,\mu)}(r) \le \mathrm{e}^{-2nr^2}, \quad r > 0.$$

In the same spirit, *expander graphs* may also satisfy some measure concentration. Let $X = (V, \mathcal{E})$ be a finite graph with set of vertices V and set of edges $\mathcal{E}$.

Assume that each vertex has at most a fixed number k_0 of adjacent edges. A subset A of V has boundary ∂A consisting of all vertices in the complement of A that are adjacent to A. Assume now that X satisfies the (linear) isoperimetric inequality

$$\mathrm{Card}(\partial A) \geq h \, \mathrm{Card}(A) \tag{2.12}$$

for some $h > 0$ and all subsets A of V such that $\mathrm{Card}(A) \leq \frac{1}{2}\mathrm{Card}(V)$. Then (2.12) amounts to the expansion property (cf. Section 1.5),

$$\mathrm{Exp}_\mu(1) \geq 1 + h > 1.$$

Here μ is a normalized counting measure on V and we give the path metric to X having all edges of unit length. By Proposition 1.13, each such expander graph has thus exponential concentration.

One class of examples are the Cayley graphs. If V is a finite group and $S \subset V$ a symmetric set of generators of V, we may join x and y in V by an edge if $x = s^{-1}y$ for some $s \in S$. Here $k_0 = \mathrm{Card}(S)$ and the path distance on $X = (V, \mathcal{E})$ is the word distance in V induced by S.

The isoperimetric inequality (2.12) is actually related to the so-called *Cheeger isoperimetric constant*, useful in Riemannian geometry (see Section 3.1). Assume for simplicity that μ is a probability measure on the Borel sets of a metric space (X, d) such that, for some constant $h > 0$ and all Borel sets A in X,

$$\mu^+(A) \geq h \min\big(\mu(A), 1 - \mu(A)\big). \tag{2.13}$$

This isoperimetric inequality is of weaker type than the spherical and Gaussian type isoperimetric inequalities. As an easy consequence of Corollary 2.2 with $v(r) = \frac{h}{2}\int_{-\infty}^{r} \mathrm{e}^{-h|x|}dx$, $r \in \mathbb{R}$, μ has an exponential concentration function.

Proposition 2.12. *Let μ be a probability measure on the Borel sets of a metric space (X, d) satisfying (2.13) for some $h > 0$. Then (X, d, μ) has exponential concentration*

$$\alpha_{(X,d,\mu)}(r) \leq \mathrm{e}^{-hr}, \quad r > 0.$$

A functional description of this result and its relation with spectrum will be provided in Section 3.1.

2.2 Brunn-Minkowski inequalities

Geometric and functional Brunn-Minkoswki inequalities may be used to provide simple but powerful concentration results.

The classical Brunn-Minkowski inequality indicates that for all bounded Borel measurable sets A, B in $\mathbb{R}^n$,

$$\mathrm{vol}_n(A + B)^{1/n} \geq \mathrm{vol}_n(A)^{1/n} + \mathrm{vol}_n(B)^{1/n} \tag{2.14}$$

where $A + B = \{x + y; x \in A, y \in B\}$ is the Minkowski sum of A and B and where we recall that $\mathrm{vol}_n(\cdot)$ denotes the volume element in $\mathbb{R}^n$. In its equivalent multiplicative form, for every $\theta \in [0, 1]$,

$$\mathrm{vol}_n\big(\theta A + (1 - \theta)B\big) \geq \mathrm{vol}_n(A)^\theta \mathrm{vol}_n(B)^{1-\theta}. \tag{2.15}$$

Indeed, under (2.14),

$$\begin{aligned}\mathrm{vol}_n\big(\theta A+(1-\theta)B\big)^{1/n} &\geq \theta\,\mathrm{vol}_n(A)^{1/n}+(1-\theta)\mathrm{vol}_n(B)^{1/n}\\ &\geq \Big(\mathrm{vol}_n(A)^{\theta}\mathrm{vol}_n(B)^{1-\theta}\Big)^{1/n}\end{aligned}$$

by the arithmetic-geometric mean inequality. Conversely, if $A'=\mathrm{vol}_n(A)^{-1/n}A$ and $B'=\mathrm{vol}_n(B)^{-1/n}B$, then (2.15) implies that $\mathrm{vol}_n(\theta A'+(1-\theta)B')\geq 1$ for every $\theta\in[0,1]$. Since

$$\theta A'+(1-\theta)B'=\frac{A+B}{\mathrm{vol}_n(A)^{1/n}+\mathrm{vol}_n(B)^{1/n}}$$

for

$$\theta=\frac{\mathrm{vol}_n(A)^{1/n}}{\mathrm{vol}_n(A)^{1/n}+\mathrm{vol}_n(B)^{1/n}},$$

(2.14) immediately follows by homogeneity.

The Brunn-Minkowski inequality may be used to produce a simple proof of Euclidean isoperimetry by taking B to be the ball with center at the origin and radius $r>0$: then (2.14) shows that

$$\mathrm{vol}_n(A_r)^{1/n}=\mathrm{vol}_n\big(A+B\big)^{1/n}\geq \mathrm{vol}_n(A)^{1/n}+v(r)^{1/n}$$

where we recall that $v(r)$ is the volume of the Euclidean ball of radius $r>0$. Since $v^{1/n}$ is linear,

$$\mathrm{vol}_n(A)^{1/n}+v(r)^{1/n}=v\Big(v^{-1}\big(\mathrm{vol}_n(A)\big)+r\Big)^{1/n}$$

so that

$$v^{-1}\big(\mathrm{vol}_n(A_r)\big)\geq v^{-1}\big(\mathrm{vol}_n(A)\big)+r$$

which amounts to isoperimetry by Proposition 2.1.

The multiplicative form of the Brunn-Minskowski inequality admits a functional version.

Theorem 2.13. *Let $\theta\in[0,1]$ and let u, v, w be non-negative measurable functions on $\mathbb{R}^n$ such that for all $x,y\in\mathbb{R}^n$,*

$$w\big(\theta x+(1-\theta)y\big)\geq u(x)^{\theta}v(y)^{1-\theta}. \tag{2.16}$$

Then

$$\int w dx\geq\Big(\int u dx\Big)^{\theta}\Big(\int v dx\Big)^{1-\theta}. \tag{2.17}$$

Applied to the characteristic functions of bounded measurable sets A and B in $\mathbb{R}^n$, it yields the multiplicative form (2.15) of the geometric Brunn-Minkowski inequality.

For the sake of completeness, we present a proof of the functional Brunn-Minkowski theorem using the transportation of measure that we learned from F. Barthe.

Proof. We start with $n = 1$ and then perform induction on dimension. By homogeneity, we may assume that $\int u dx = \int v dx = 1$ and by approximation that u and v are continuous with strictly positive values. Define $x, y : [0,1] \to \mathbb{R}$ by

$$\int_{-\infty}^{x(t)} u(q)dq = t, \quad \int_{-\infty}^{y(t)} v(q)dq = t.$$

Therefore x and y are increasing and differentiable and

$$x'(t)u\big(x(t)\big) = y'(t)v\big(y(t)\big) = 1.$$

Set $z(t) = \theta x(t) + (1-\theta)y(t)$, $t \in [0,1]$. By the arithmetic-geometric mean inequality, for every t,

$$z'(t) = \theta x'(t) + (1-\theta)y'(t) \geq \big(x'(t)\big)^{\theta}\big(y'(t)\big)^{1-\theta}. \tag{2.18}$$

Now, since z is injective, by the hypothesis (2.16) on w and (2.18),

$$\begin{aligned}
\int w dx &\geq \int_0^1 w\big(z(t)\big)z'(t)dt \\
&\geq \int_0^1 u\big(x(t)\big)^{\theta} v\big(y(t)\big)^{1-\theta} \big(x'(t)\big)^{\theta}\big(y'(t)\big)^{1-\theta} dt \\
&= \int_0^1 \big[u\big(x(t)\big)x'(t)\big]^{\theta}\big[v\big(y(t)\big)y'(t)\big]^{1-\theta} dt \\
&= 1.
\end{aligned}$$

This proves the case $n = 1$. It is then easy to deduce the general case by induction on n. Suppose $n > 1$ and assume the Brunn-Minkowski theorem holds in $\mathbb{R}^{n-1}$. Let u, v, w be non-negative measurable functions on $\mathbb{R}^n$ satisfying (2.16) for some $\theta \in [0,1]$. Let $q \in \mathbb{R}$ be fixed and define $u_q : \mathbb{R}^{n-1} \to [0,\infty)$ by $u_q(x) = u(x,q)$ and similarly for v_q and w_q. Clearly, if $q = \theta q_0 + (1-\theta)q_1$, $q_0, q_1 \in \mathbb{R}$,

$$w_q\big(\theta x + (1-\theta)y\big) \geq u_q(x)^{\theta} v_q(y)^{1-\theta}$$

for all $x, y \in \mathbb{R}^{n-1}$. Therefore, by the induction hypothesis,

$$\int_{\mathbb{R}^{n-1}} w_q dx \geq \left(\int_{\mathbb{R}^{n-1}} u_q dx\right)^{\theta} \left(\int_{\mathbb{R}^{n-1}} v_q dx\right)^{1-\theta}.$$

Finally, applying the one-dimensional case shows that

$$\int w dx = \int \left(\int_{\mathbb{R}^{n-1}} w_q dx\right) dq \geq \left(\int u dx\right)^{\theta} \left(\int v dx\right)^{1-\theta}$$

which is the desired result. Theorem 2.13 is established. □

Brunn-Minkowski inequalities may be used to produce concentration type inequalities, and to recover in particular the application of Theorem 2.7 to concentration. We however start with a milder result for log-concave measures.

On $\mathbb{R}^n$ (for simplicity), we say that a Borel probability measure μ is *log-concave* if for any Borel sets A and B in $\mathbb{R}^n$, and every $\theta \in [0,1]$,

$$\mu\big(\theta A + (1-\theta)B\big) \geq \mu(A)^\theta \mu(B)^{1-\theta}. \tag{2.19}$$

If μ has strictly positive density e^{-U}, μ is log-concave if and only if U is convex. If $\mathcal{K}$ is any convex body of finite volume in $\mathbb{R}^n$, the uniform normalized measure on $\mathcal{K}$ defined by

$$\mu_{\mathcal{K}}(A) = \frac{\mathrm{vol}_n(A \cap \mathcal{K})}{\mathrm{vol}_n(\mathcal{K})} \tag{2.20}$$

is log-concave by the Brunn-Minkowski inequality.

The following is a concentration result for homothetics.

Proposition 2.14. *Let μ be log-concave. Then, for all convex symmetric sets A in $\mathbb{R}^n$ with measure $\mu(A) = a > 0$,*

$$1 - \mu(rA) \leq a\Big(\frac{1-a}{a}\Big)^{(r+1)/2}$$

for every $r > 1$. In particular, if $a > \frac{1}{2}$,

$$\limsup_{r \to \infty} \frac{1}{r} \log\big(1 - \mu(rA)\big) < 0.$$

Proof. Since A is convex and symmetric, one may check that

$$\mathbb{R}^n \setminus A \supset \frac{2}{r+1}\big(\mathbb{R}^n \setminus (rA)\big) + \frac{r-1}{r+1} A.$$

Then, by (2.19),

$$1 - \mu(A) \geq \big(1 - \mu(rA)\big)^{2/(r+1)} \mu(A)^{(r-1)/(r+1)}$$

from which the result follows. □

If F on $\mathbb{R}^n$ is non-negative, convex and symmetric, and if $A = \{|F| \leq m\}$ with $\mu(A) \geq \frac{3}{4}$ for example, then we get from Proposition 2.14 that for every $r > 1$,

$$\mu\big(\{|F| \geq mr\}\big) \leq \mathrm{e}^{-r/2}.$$

Integrating in r shows that for some numerical constant $C > 0$, $\|F\|_q \leq Cqm$ for every $q \geq 1$ where $\|\cdot\|_q$ is the L^q-norm with respect to μ. Since we may choose $m = 4\|F\|_1$, it follows that, for some $C > 0$ and all $q \geq 1$,

$$\|F\|_q \leq Cq\,\|F\|_1. \tag{2.21}$$

This result applies in particular to the normalized Lebesgue measure $\mu_{\mathcal{K}}$ (2.20) on a convex body $\mathcal{K}$. The reverse Hölder inequality (2.21) is rather useful in the geometry of convex bodies. For example, if $\mathcal{K} = [-\frac{1}{2}, +\frac{1}{2}]^n$ and $F(x) = \|\sum_{i=1}^n x_i v_i\|$, $x =$

$(x_1, \dots, x_n) \in \mathbb{R}^n$, for vectors $v_1, \dots, v_n$ in some normed vector space $(E, \|\cdot\|)$, the preceding yields

$$\left(\int_{[-\frac{1}{2},+\frac{1}{2}]^n} \left\|\sum_{i=1}^n x_i v_i\right\|^q dx\right)^{1/q} \leq C\,q \int_{[-\frac{1}{2},+\frac{1}{2}]^n} \left\|\sum_{i=1}^n x_i v_i\right\| dx. \tag{2.22}$$

These inequalities are part of the Khintchine-Kahane inequalities (see Section 7.1).

We now consider strict log-concavity conditions. Assume indeed that μ is a probability measure on $\mathbb{R}^n$ with smooth strictly positive density e^{-U} with respect to Lebesgue measure such that for some $c > 0$ and all $x, y \in \mathbb{R}^n$,

$$U(x) + U(y) - 2U\Big(\frac{x+y}{2}\Big) \geq \frac{c}{4}\,|x-y|^2. \tag{2.23}$$

A typical example is of course the canonical Gaussian measure γ on $\mathbb{R}^n$ for which $c = 1$. Given a bounded measurable function f on $\mathbb{R}^n$, apply then the functional Brunn-Minkowski Theorem 2.13 to

$$u(x) = \mathrm{e}^{Q_{c/2}f(x) - U(x)}, \quad v(y) = \mathrm{e}^{-f(y)-U(y)}, \quad w(z) = \mathrm{e}^{-U(z)},$$

where we define, following (1.27), $Q_c f$, $c > 0$, as the infimum-convolution

$$Q_c f(x) = \inf_{y\in\mathbb{R}^n}\Big[f(y) + \frac{c}{2}\,|x-y|^2\Big], \quad x \in \mathbb{R}^n,$$

with cost $\tilde{c}(x,y) = \frac{c}{2}\,|x-y|^2$, $(x,y) \in \mathbb{R}^n \times \mathbb{R}^n$. By definition of $Q_c f$ and the convexity hypothesis (2.23) on U, condition (2.16) is satisfied with $\theta = \frac{1}{2}$ so that

$$1 \geq \int \mathrm{e}^{Q_{c/2}f} d\mu \int \mathrm{e}^{-f} d\mu.$$

We then make use of Proposition 1.18 to obtain a first alternative argument for the concentration results of Corollary 2.6 and Theorem 2.7 which were deduced there from isoperimetry.

Theorem 2.15. *Let $d\mu = \mathrm{e}^{-U}dx$ where U satisfies (2.23). Then,*

$$\alpha_\mu(r) \leq 2\,\mathrm{e}^{-cr^2/4}, \quad r > 0.$$

In particular

$$\alpha_\gamma(r) \leq 2\,\mathrm{e}^{-r^2/4}, \quad r > 0,$$

for the canonical Gaussian measure γ on $\mathbb{R}^n$.

Note that the constants provided by Theorem 2.15 are not quite optimal. Theorem 2.15 applies similarly to arbitrary norms $\|\cdot\|$ on $\mathbb{R}^n$ instead of the Euclidean norm in (2.23). Concentration then takes place with respect to the metric given by the norm. We may also work with p-convex potentials U, $p \geq 2$, satisfying for some $c > 0$ and all $x, y \in \mathbb{R}^n$,

$$U(x) + U(y) - 2U\Big(\frac{x+y}{2}\Big) \geq \frac{c}{2p}\,\|x-y\|^p, \tag{2.24}$$

to produce concentration of the type

$$\alpha_\mu(r) \le 2\,\mathrm{e}^{-c_p r^p}, \quad r > 0,$$

for some $c_p > 0$ only depending on $p \ge 2$.

In the last part of this section, we discuss how Brunn-Minkowski inequalities may be further used to determine concentration properties for uniform and surface measures on the unit ball of uniformy convex spaces, extending the case of the standard sphere.

A normed space $X = (\mathbb{R}^n, \|\cdot\|)$ is said to be *uniformly convex* if for each $\varepsilon > 0$ there is $\delta = \delta(\varepsilon) > 0$ such that whenever $x, y \in X$ with $\|x\| = \|y\| = 1$ and $\|x - y\| \ge \varepsilon$, then

$$1 - \frac{\|x+y\|}{2} \ge \delta(\varepsilon).$$

The *modulus of convexity* of the normed space X is thus defined as the function of ε, $0 < \varepsilon \le 2$,

$$\delta_X(\varepsilon) = \inf\left\{1 - \left\|\frac{x+y}{2}\right\|; \|x\| = \|y\| = 1, \|x-y\| \ge \varepsilon\right\}.$$

L^p-spaces, $1 < p < \infty$, are uniformly convex with modulus of convexity of power type, $\delta(\varepsilon) \ge C\varepsilon^{\max(p,2)}$ for every ε. Further examples are discussed in [Li-T].

Now let X be uniformly convex and let $\mathcal{B}$ be the unit ball of $(X, \|\cdot\|)$. Denote by $\mu_{\mathcal{B}}$ the normalized uniform volume element on $\mathcal{B}$. Given two non-empty sets $A, B \subset \mathcal{B}$ at distance $\varepsilon \in (0,1)$, we have by definition of the modulus of convexity that

$$\frac{1}{2}\,(A+B) \subset \big(1 - \delta_X(\varepsilon)\big)\mathcal{B}.$$

Hence, by the Brunn-Minkowski inequality (2.15),

$$\mu_{\mathcal{B}}^{1/2}(A)\,\mu_{\mathcal{B}}^{1/2}(B) \le \big(1 - \delta_X(\varepsilon)\big)^n.$$

Taking for B the complement of A_ε, we get,

$$\begin{aligned} \mu_{\mathcal{B}}(A_\varepsilon) &\ge 1 - \frac{1}{\mu_{\mathcal{B}}(A)}\,\big(1 - \delta_X(\varepsilon)\big)^{2n} \\ &\ge 1 - \frac{1}{\mu_{\mathcal{B}}(A)}\,\mathrm{e}^{-2n\delta_X(\varepsilon)}. \end{aligned} \tag{2.25}$$

Here we recover in particular concentration on the Euclidean unit ball in $\mathbb{R}^n$ (Proposition 2.9).

This measure concentration result for $\mu_{\mathcal{B}}$ may easily be transferred to the normalized surface measure $\mu_{\partial\mathcal{B}}$ of $\mathcal{B}$ with respect to itself, defined by

$$\mu_{\partial\mathcal{B}}(A) = \mu_{\mathcal{B}}\bigg(\bigcup_{0\le t\le 1} tA\bigg), \quad A \subset \partial\mathcal{B}. \tag{2.26}$$

Indeed, given a measurable set A on $\partial\mathcal{B}$, consider

$$\Gamma = \bigcup_{1/2\le t\le 1} tA \subset \mathcal{B}.$$

Clearly $\mu_{\mathcal{B}}(\Gamma) \geq \frac{1}{2}\,\mu_{\partial\mathcal{B}}(A)$ while $\Gamma_\varepsilon \subset \bigcup_{0\leq t\leq 1} tA_{2\varepsilon}$. Therefore, as a consequence of (2.25), we obtain the following result that covers in particular the standard sphere $\mathbb{S}^n$ itself.

Theorem 2.16. *Denote by $\mu_{\partial\mathcal{B}}$ the surface measure as defined in (2.26) on the boundary of the unit ball $\mathcal{B}$ of a uniformly convex space $X = (\mathbb{R}^n, \|\cdot\|)$ with modulus of convexity δ_X. Then,*

$$\alpha_{(X,\|\cdot\|,\mu_{\partial\mathcal{B}})}(r) \leq 4\,\mathrm{e}^{-2n\delta_X(r/2)}, \quad r > 0.$$

As we have seen, $\delta(\varepsilon) \geq C\varepsilon^{\max(p,2)}$ on L^p-spaces, $1 < p < \infty$. Therefore, if $\mathcal{B}_p^n$ is the ball in $\mathbb{R}^n$ for the norm

$$|x|_p = \left(\sum_{i=1}^n |x_i|^p\right)^{1/p}, \quad x = (x_1, \ldots, x_n) \in \mathbb{R}^n,$$

and if μ is normalized Lebesgue measure on $\mathcal{B}_p^n$, or its boundary, then

$$\alpha_{(\mathcal{B}_p^n,|\cdot|_p,\mu)}(r) \leq C\mathrm{e}^{-cnr^p}, \quad r > 0,$$

for $2 \leq p < \infty$ and

$$\mathrm{ObsDiam}_\mu\big(\mathcal{B}_p^n, |\cdot|_p\big) = O\big(n^{-1/p}\big),$$

while for $1 < p \leq 2$,

$$\alpha_{(\mathcal{B}_p^n,|\cdot|_p,\mu)}(r) \leq C\mathrm{e}^{-cnr^2}, \quad r > 0, \tag{2.27}$$

and

$$\mathrm{ObsDiam}_\mu\big(\mathcal{B}_p^n, |\cdot|_p\big) = O\big(n^{-1/2}\big).$$

The functions $x \mapsto |x|_p^p$ are examples of p-convex potentials, $p \geq 2$, in the sense of (2.24) (or more generally $x \mapsto \|x\|^p$ for a norm $\|\cdot\|$ with modulus of convexity of power type $p \geq 2$; cf. [Li-T]). Therefore, considering the product probability measure on $\mathbb{R}^n$ when each coordinate is endowed with the distribution $c_p\,\mathrm{e}^{-|x|^p}dx$, the preceding may also be deduced from the analogues of Propositions 2.9 and 2.10 for this measure. This is the argument outlined at the end of Chapter 4 to prove that (2.27) also holds in this form for $p = 1$.

2.3 Semigroup tools

In this section, we present some rather elementary semigroup arguments to reach the full concentration properties on spheres and manifolds with strictly positive curvature as well as for Gaussian measures and strictly log-concave distributions derived in Section 2.1 from sharp isoperimetric inequalities. The functional approach we adopt, relying only on Bochner's formula, allows a neat and tireless treatment of the geometric Ricci curvature.

We start with the case of the sphere, or more generally of a Riemannian manifold with a strictly positive lower bound on the Ricci curvature. We then describe how the formal argument extends. Thus, as in Section 2.1, let (X, g) be a compact connected smooth Riemannian manifold of dimension n (≥ 2) equipped with the normalized Riemannian volume element $d\mu = \frac{dv}{V}$ where V is the total volume of X. Denote by $c = c(X)$ the infimum of the Ricci curvature tensor over all unit tangent vectors, and assume that $c > 0$. We use Laplace's bounds to establish the following result. By Proposition 1.7, it shows that (X, g, μ) has normal concentration

$$\alpha_{(X,g,\mu)}(r) \leq 2\,\mathrm{e}^{-cr^2/8}, \quad r > 0.$$

Proposition 2.17. *Let (X, g) be a compact connected smooth Riemannian manifold of dimension n (≥ 2) equipped with the normalized Riemannian volume element $d\mu = \frac{dv}{V}$ such that $c(X) > 0$. For any 1-Lipschitz function F on X, and any $r \geq 0$,*

$$\mu\big(\{F \geq \textstyle\int F d\mu + r\}\big) \leq \mathrm{e}^{-cr^2/2}.$$

Proof. By Proposition 1.14, it is enough to show that

$$\mathrm{E}_{(X,g,\mu)}(\lambda) \leq \mathrm{e}^{\lambda^2/2c} \tag{2.28}$$

for every $\lambda \geq 0$ where we recall that $\mathrm{E}_{(X,g,\mu)}$ is the Laplace functional of μ defined in Section 1.6.

Let ∇ be the gradient on (X, g) and Δ be the Laplace-Beltrami operator. *Bochner's formula* (see [G-H-L], [Cha2], etc.) indicates that for every smooth function f on X,

$$\frac{1}{2}\,\Delta\big(|\nabla f|^2\big) - \nabla f \cdot \nabla(\Delta f) = \|\mathrm{Hess} f\|_2^2 + \mathrm{Ric}(\nabla f, \nabla f) \tag{2.29}$$

where $\|\mathrm{Hess} f\|_2$ is the Hilbert-Schmidt norm of the Hessian of f and Ric is the Ricci tensor. By the hypothesis,

$$\frac{1}{2}\,\Delta\big(|\nabla f|^2\big) - \nabla f \cdot \nabla(\Delta f) \geq c\,|\nabla f|^2. \tag{2.30}$$

This is the only place where Ricci curvature is really used for the concentration result of Proposition 2.17. We now take (2.30) into account by functional semigroup tools.

Consider the heat semigroup $P_t = \mathrm{e}^{t\Delta}$, $t \geq 0$, with generator the Laplace operator Δ on X (cf. [Yo], [Davie2], [F-O-T], etc.). Given a (regular) real-valued function f on X,

$$u = u(x, t) = P_t f(x), \quad x \in X,\, t \geq 0,$$

is the solution of the initial value problem for the partial differential (heat) equation

$$\frac{\partial u}{\partial t} = \Delta u \tag{2.31}$$

with $u(\cdot, 0) = f$.

Let f be smooth on X. It is an immediate consequence of (2.30) that, for every $t \geq 0$,

$$|\nabla P_t f|^2 \leq \mathrm{e}^{-2ct} P_t\big(|\nabla f|^2\big). \tag{2.32}$$

Indeed, let $t \geq 0$ be fixed and set $\psi(s) = \mathrm{e}^{-2cs} P_s(|\nabla P_{t-s} f|^2)$ for every $0 \leq s \leq t$. By (2.31),

$$\psi'(s) = 2\mathrm{e}^{-2cs}\Big[-cP_s\big(|\nabla P_{t-s}f|^2\big) \\ + P_s\big(\tfrac{1}{2}\,\Delta\big(|\nabla P_{t-s}f|^2\big) - \nabla P_{t-s}f \cdot \nabla(\Delta P_{t-s}f)\big)\Big].$$

Hence, by (2.30) applied to $P_{t-s}f$ for every s, ψ is non-decreasing and (2.32) follows.

Now, let F be a mean zero 1-Lipschitz function on (X, g) and let $\lambda \geq 0$. We may and do assume that F is smooth enough for what follows. In particular $|\nabla F| \leq 1$ almost everywhere. Set $\Psi(t) = \int \mathrm{e}^{\lambda P_t F} d\mu$, $t \geq 0$. Since $\Psi(\infty) = 1$ and since, by (2.32),

$$|\nabla P_t F|^2 \leq \mathrm{e}^{-2ct},$$

we can write for every $t \geq 0$,

$$\begin{aligned}
\Psi(t) &= 1 - \int_t^\infty \Psi'(s)\,ds \\
&= 1 - \lambda \int_t^\infty \Big(\int \Delta(P_s F)\mathrm{e}^{\lambda P_s F}\,d\mu\Big)ds \\
&= 1 + \lambda^2 \int_t^\infty \Big(\int \big|\nabla P_s F\big|^2 \mathrm{e}^{\lambda P_s F}\,d\mu\Big)ds \\
&\leq 1 + \lambda^2 \int_t^\infty \mathrm{e}^{-2cs}\,\Psi(s)\,ds
\end{aligned}$$

where we used integration by parts in the space variable. By Gronwall's lemma,

$$\Psi(0) = \int \mathrm{e}^{\lambda F} d\mu \leq \mathrm{e}^{\lambda^2/2c}$$

which is (2.28), at least for a smooth function F. The general case follows from a standard regularization procedure. The proof is complete. □

The principle of proof of Proposition 2.17 applies similarly to further examples of analytic interest, which include the concentration result of Theorem 2.7 and provide an alternative description of the conclusion of Theorem 2.15, which was based on the Brunn-Minkowski approach. We only outline the argument. Consider indeed, a Borel probability measure μ on $\mathbb{R}^n$, with density e^{-U} with respect to Lebesgue measure, where U is a smooth function on $\mathbb{R}^n$. On U we adopt some convexity assumption that corresponds to the Ricci curvature lower bound in a Riemannian context. Suppose indeed that, as in Theorem 2.7, for some $c > 0$, $\mathrm{Hess}\, U(x) \geq c\,\mathrm{Id}$ uniformly in $x \in \mathbb{R}^n$. A typical example is the Gaussian density $U(x) = |x|^2/2$ (up to the normalization factor) for which $c = 1$. Consider then the second order differential operator $\mathrm{L} = \Delta - \nabla U \cdot \nabla$ on $\mathbb{R}^n$ with invariant and symmetric measure μ. Integration by parts for L reads as

$$\int f(-\mathrm{L}g)d\mu = \int \nabla f \cdot \nabla g\, d\mu \tag{2.33}$$

for smooth f, g. It is then a mere exercise to check the analogues of Bochner's formula (2.29) and of (2.30), namely

$$\frac{1}{2}\,\mathrm{L}(|\nabla f|^2) - \nabla f \cdot \nabla(\mathrm{L}f) = \|\mathrm{Hess} f\|_2^2 + \mathrm{Hess}\,(U)(\nabla f, \nabla f)$$

and, by the convexity assumption on U,

$$\frac{1}{2}\,\mathrm{L}(|\nabla f|^2) - \nabla f \cdot \nabla(\mathrm{L}f) \geq c\,|\nabla f|^2. \tag{2.34}$$

Following then the standard Hille-Yoshida theory of self-adjoint operators [Yo], we may consider, under mild growth conditions on U, the invariant and time reversible semigroup $(P_t)_{t\geq 0}$ with infinitesimal generator L (cf. also [Du-S], [Fr], [F-O-T], [Davie1], [Roy], etc., and below for a probabilistic outline). Therefore, in particular, $\frac{\partial P_t f}{\partial t} = \mathrm{L}P_t f$. Strict convexity of U easily enters this framework. For example, in the case of the canonical Gaussian measure γ on $\mathbb{R}^n$, $\mathrm{L}f(x) = \Delta f(x) - x \cdot \nabla f(x)$, $x \in \mathbb{R}^n$, for f smooth enough, and the semigroup $(P_t)_{t\geq 0}$ with generator L admits the integral representation

$$P_t f(x) = \int_{\mathbb{R}^n} f\big(\mathrm{e}^{-t}x + (1 - \mathrm{e}^{-2t})^{1/2} y\big)\, d\gamma(y), \quad x \in \mathbb{R}^n,\, t \geq 0.$$

Now (2.34) may be used exactly in the same way to show the corresponding commutation inequality (2.32) for $(P_t)_{t\geq 0}$. The same proof as the one of Proposition 2.17 thus leads to the following result.

Proposition 2.18. *Let $d\mu = \mathrm{e}^{-U}dx$ be a probability measure on the Borel sets of $\mathbb{R}^n$ such that, for some $c > 0$, $\mathrm{Hess}\, U(x) \geq c\,\mathrm{Id}$ uniformly in $x \in \mathbb{R}^n$. Then, for every bounded 1-Lipschitz function $F : \mathbb{R}^n \to \mathbb{R}$ and every $r \geq 0$,*

$$\mu\big(\{F \geq \textstyle\int F d\mu + r\}\big) \leq \mathrm{e}^{-cr^2/2}.$$

Together with Proposition 1.7, we thus find again the concentration properties of μ derived from isoperimetry in Theorem 2.7 or from Brunn-Minskowski inequalities in Theorem 2.15.

As a consequence, if γ is the canonical Gaussian measure on $\mathbb{R}^n$ with density $(2\pi)^{-n/2}\mathrm{e}^{-|x|^2/2}$ with respect to Lebesgue measure, for every 1-Lipschitz function F on $\mathbb{R}^n$ and every $r \geq 0$,

$$\gamma\big(\{F \geq \textstyle\int F d\gamma + r\}\big) \leq \mathrm{e}^{-r^2/2}, \quad r \geq 0. \tag{2.35}$$

This inequality must be compared with the corresponding inequality for the median in (2.10). It extends similarly the case of linear functionals.

An alternative proof of (2.35) may be provided by a beautiful simple argument using Brownian integration. For every smooth function f on $\mathbb{R}^n$, write

$$f\big(W(1)\big) - \mathbb{E} f\big(W(1)\big) = \int_0^1 \nabla P_{1-t} f\big(W(t)\big) \cdot dW(t) \tag{2.36}$$

where $(W(t))_{t\geq 0}$ is Brownian motion on $\mathbb{R}^n$ starting at the origin defined on some probability space $(\Omega, \mathcal{F}, \mathbb{P})$ and where $(P_t)_{t\geq 0}$ denotes its associated semigroup (the heat semigroup), with the probabilistic normalization (that is, with generator $\frac{1}{2}\Delta$). Note then that the above stochastic integral has the same distribution as $\beta(T)$ where $(\beta(t))_{t\geq 0}$ is a one-dimensional Brownian motion and where

$$T = \int_0^1 \left|\nabla P_{1-t} f(W(t))\right|^2 dt.$$

Therefore, for every 1-Lipschitz function F, $T \leq 1$ almost surely so that, for all $r \geq 0$,

$$\begin{aligned}\mathbb{P}\big(\{F(W(1)) - \mathbb{E}F(W(1)) \geq r\}\big) &\leq \mathbb{P}\big(\{\max_{0\leq t\leq 1} \beta(t) \geq r\}\big) \\ &= 2\int_r^\infty \mathrm{e}^{-x^2/2}\,\frac{dx}{\sqrt{2\pi}} \\ &\leq \mathrm{e}^{-r^2/2}.\end{aligned}$$

Since $W(1)$ has distribution γ, this is thus simply (2.35).

As time of Brownian motion changes, continuous martingales satisfy similarly sharp deviation inequalities. Assume for simplicity that $M = (M_t)_{0\leq t\leq 1}$ is a real-valued continuous martingale on some filtered probability space $(\Omega, \mathcal{F}, \mathcal{F}_t, \mathbb{P})$ satisfying the usual conditions (cf. [R-Y]). Denote by $(\langle M, M\rangle_t)_{0\leq t\leq 1}$ its quadratic increasing process and assume that $\langle M\rangle_1 \leq D^2$ almost surely. Then, for every $r \geq 0$,

$$\begin{aligned}\mathbb{P}\big(\{M_1 \geq \mathbb{E}(M_1) + r\}\big) &\leq \mathbb{P}\big(\{\max_{0\leq t\leq \langle M,M\rangle_1} \beta(t) \geq r\}\big) \\ &\leq \mathbb{P}\big(\{\max_{0\leq t\leq D^2} \beta(t) \geq r\}\big)\end{aligned}$$

so that

$$\mathbb{P}\big(\{M_1 \geq \mathbb{E}(M_1) + r\}\big) \leq \mathrm{e}^{-r^2/2D^2}. \tag{2.37}$$

Alternatively, and assuming that $M_0 = 0$ for simplicity,

$$\mathrm{e}^{\lambda M_t - \frac{\lambda^2}{2}\langle M,M\rangle_t}, \quad \lambda \in \mathbb{R},\ t \geq 0,$$

is a martingale with expectation 1. Hence, for any $\lambda, r \geq 0$,

$$\begin{aligned}\mathbb{P}\big(\{M_1 \geq r\}\big) &\leq \mathbb{P}\big(\{\lambda M_1 - \tfrac{\lambda^2}{2}\langle M, M\rangle_1 \geq \lambda r - \tfrac{\lambda^2}{2} D^2\}\big) \\ &\leq \mathrm{e}^{-\lambda r + \lambda^2 D^2/2}\end{aligned}$$

from which we recover (2.37) by optimizing in λ.

Related to the representation formula (2.36), a somewhat analogous method may be developed in the setting of general Markov processes under the technique of *forward and backward martingales*. For simplicity, suppose that $(\xi_t)_{t\geq 0}$ is a strong Markov process on some probability space with continuous paths taking its values in a locally compact space X with countable base, that it is time homogeneous, and that for each $x \in X$, there is a probability measure $\mathbb{P}^x$ on $\Omega = C([0,\infty), X)$ corresponding to the law of ξ conditioned to start with $\xi_0 = x$. More generally, let $\mathbb{P}^\mu$ be the law of ξ where the law of ξ_0 is a Radon measure μ. Such a process induces a Markov semigroup $(P_t)_{t\geq 0}$ on bounded Borel functions defined by

$$P_t f(x) = \mathbb{E}\big(f(\xi_t) \,|\, \xi_0 = x\big) = \mathbb{E}^x\big(f(\xi_t)\big), \quad t \geq 0,\, x \in X.$$

Because of the Markov property, $P_s P_t = P_{s+t}$.

Suppose for regularity purposes that $P_t : C_b(X) \to C_b(X)$. If $(P_t)_{t\geq 0}$ has stationary (not necessarily finite) measure μ (that is, $\mathbb{P}^\mu(\{X_t \in A\}) = \mu(A)$ for all t or equivalently $\int P_t f d\mu = \int f d\mu$), then P_t maps $\mathrm{L}^p(\mu)$ into itself for every $1 \leq p \leq \infty$. We can thus express $P_t = \mathrm{e}^{t\mathrm{L}}$, where L is a uniquely defined closed (unbounded) operator from $\mathrm{L}^p \to \mathrm{L}^p$ as well as from $C_b(X) \to C_b(X)$ known as the infinitesimal generator of ξ (cf. [Yo], [S-V], [I-W], [F-O-T]).

Now let $f \in \mathcal{D}(\mathrm{L}) \cap C_b(X)$ where $\mathcal{D}(\mathrm{L})$ is the domain of L. Itô's formula of martingale characterization expresses that

$$f(\xi_t) - f(\xi_0) = M_t + \int_0^t \mathrm{L} f(\xi_s) ds \tag{2.38}$$

where $M_t = M_t^{f,\xi}$ is a $\mathbb{P}^\mu$-continuous local martingale whenever $\mathrm{L}f$ is in $\mathrm{L}^1(\mu)$ and a $\mathbb{P}^x$-local martingale for any f satisfying $\mathrm{L}f \in C_b(X)$. In concrete situations, the operator L is frequently the closure of a second order elliptic operator. This is the situation we encountered above with the Laplace-Beltrami operator on a manifold (X, g). In these geometric cases, one can usually identify the quadratic variation process $\langle M, M\rangle_t$ of M as

$$\langle M, M\rangle_t = \int_0^t \big|\nabla f(\xi_s)\big|^2 ds.$$

When the operator L is selfadjoint, the adjoint semigroup is the same as the original one. This analytic remark has stochastic counterparts. The processes ξ_{1-t} and ξ_t are identical in law when started with initial distribution μ. We say that the process is reversible with respect to μ. In particular, $M_t^{f,\xi_{1-t}} = \widetilde{M}_t$ is also a martingale (in the filtration $\mathcal{F}_t = \sigma(\xi_{1-s}, s < t)$). So, by (2.38),

$$f(\xi_t) = f(\xi_0) + M_t + \int_0^t \mathrm{L} f(\xi_s) ds$$

and

$$f(\xi_t) = f(\xi_1) + \widetilde{M}_t + \int_0^{1-t} \mathrm{L} f(\xi_{1-s}) ds.$$

Taking the two expressions together,

$$2f(\xi_t) = \big(f(\xi_0) + f(\xi_1)\big) + \big(M_t + \widetilde{M}_{1-t}\big) + \int_0^1 \mathrm{L} f(\xi_s) ds,$$

and thus

$$df(\xi_t) = \frac{1}{2}\big(dM_t + d\widetilde{M}_{1-t}\big).$$

Here the most useful form is

$$f(\xi_1) - f(\xi_0) = M_1 - \widetilde{M}_1$$

where M and $\widetilde{M}$ are both martingales with

$$d\langle M, M\rangle_t = d\langle \widetilde{M}, \widetilde{M}\rangle_{1-t} = \left|\nabla f(\xi_t)\right|^2 dt.$$

It then follows from (2.37) that whenever F is 1-Lipschitz, for every $r \geq 0$,

$$\mathbb{P}^\mu\big(\big\{|F(\xi_1) - F(\xi_0)| \geq r\big\}\big) \leq 2\,\mathrm{e}^{-r^2/8}.$$

To make use of this result, taking, for example, F to be the distance function to some fixed set A shows that

$$\mathbb{P}^\mu\big(\big\{\xi_0 \in A, d(\xi_1, A) \geq r\big\}\big) \leq 2\,\mathrm{e}^{-r^2/8}$$

for every $r \geq 0$. While such a bound does not require any geometric assumptions on curvature as in previous analogous results, its application is conditioned by the initial information $\xi_0 \in A$ that is handled by geometric arguments involving volume of balls and curvature. Further applications in probability and potential theory are discussed in [L-Z], [Tak], [Ly].

Notes and Remarks

The isoperimetric inequality on the sphere goes back to P. Lévy [Lé] and E. Schmidt [Schmi]. A short self-contained proof is presented in the appendix of [F-L-M]. An alternative method relying on two-point symmetrization is used in [B-T] and presented in [Beny], [Sche5]. Isoperimetric inequalities are discussed in [B-Z], [Os], etc. Lévy's original idea [Lé] was extended by M. Gromov [Grom1] (see [M-S], [Grom2], [G-H-L]) as a comparison theorem for Riemannian manifolds with a strictly positive lower bound on the Ricci curvature (Theorem 2.4). Further geometric applications of Theorem 2.4 are discussed in [Grom2]. Concentration for Grassmann and Stiefel manifolds, orthogonal groups, and first spectrum applications go back to the pioneering works [Mi1], [Mi3] by V. Milman. The Gaussian isoperimetric inequality (Theorem 2.6) is due independently to C. Borell [Bor2] and V. N. Sudakov and B. Tsirel'son [S-T] as limiting spherical isoperimetry. An intrinsic proof using Gaussian symmetrization is due to A. Ehrhard [Eh] while S. Bobkov gave in [Bob4] a functional argument based on a two-point inequality and the central limit theorem. The paper [Bor2] develops the infinite dimensional aspects of the Gaussian isoperimetric inequality (see [Li], [Bog], and for applications, [Le3]). Theorem 2.7 is established in a more general context in [Ba-L] via the functional formulation of [Bob4] (see also [Bob3]) with the semigroup tools of Section 2.3. This result may also be proved more simply by the localization lemma of L. Lovász and M. Simonovits [L-S] as developed in [K-L-S] and explained in [Grom2] (using the separation ideas of [Gr-M2]). See also [Ale]. Proposition 2.8 was observed in [I-S-T] and [Pis1] and Proposition 2.10 in [M-S] (see also [Grom2]). Theorem 2.11 is a direct consequence of the solution of the isoperimetric problem on the discrete cube by L. H. Harper [Ha] (for a simple proof see [F-F] and for a far reaching generalization [W-W]).

Historical aspects of the Brunn-Minkowski theorem are discussed in [DG]. To its functional version, the names of R. Henstock and A. MacBeath, A. Prékopa, L. Leindler, H. Brascamp and E. Lieb, C. Borell, etc. should be associated. Pertinence of Brunn-Minkowski inequalities to measure concentration has a long run (see [M-S]). Proposition 2.14 is due to C. Borell [Bor1]. Improved (and optimal) estimates are obtained in [L-S] by means of the localization method for the uniform distribution on the Euclidean ball and extended in [Gu] to all log-concave measures. Inequality (2.21) is fruitful in the analysis of high dimensional convex sets and was extended to multilinear functions in [Bou] (see also [Bob7]). Recent developments on levels of concentration with respect to a given class of functions with new geometric applications to concentration of random sets in $\mathbb{R}^n$ with respect to linear functions are initiated in [Gi-M2] (see also [Mi7]). For a recent application of concentration to the central limit theorem for convex bodies, see [A-B-P]. Inequality (2.22) is part of the Khintchine-Kahane inequalities [Ka] and will be investigated more precisely in Section 7.1.

Cheeger type isoperimetric inequalities for log-concave measures are studied in [Bob5]. The proof of Theorem 2.15 is essentially due to B. Maurey [Mau2] (see also [Schmu1], [B-G], [Bo-L3]). Extensions of the functional Brunn-Minkowski theorem to Riemannian manifolds were described recently in [CE-MC-S] and may be used to recover the concentration results of Theorem 2.4. Theorem 2.16 is due to M. Gromov and V. Milman [Gr-M2], the simple proof taken from [AR-B-V]. See also [Schmu1], [Bo-L3]. More on concentration in L^p-spaces may be found in [Sche3], [Sche5], [S-Z1], [S-S1] (see also Chapter 4).

The semigroup arguments of Section 2.3 were developed originally for logarithmic Sobolev inequalities by D. Bakry and M. Emery [B-E] (see [Bak1], [Bak2]) and their adaptation to concentration inequalities was noticed in [Le1]. Propositions 2.17 and 2.18 are taken from [Le1] (see also [Le5]). General diffusions and Lipschitz functions with respect to the carré du champ operator are treated similarly [Bak1], [Le6]. The discrete analogue for non-local generators on graphs is studied in [Schmu3] by appropriate notion of curvature of a graph (with illustrations).

Further semigroup arguments relying on logarithmic Sobolev inequalities and covariance identities are developed in the same spirit in Chapter 5.

The Brownian proof of (2.35) is due to I. Ibragimov, B. Tsirel'son and V. Sudakov [I-T-S], and was unfortunately ignored for a long time. The representation formula (2.36) is a particular form of the important Clark-Ocone formula in stochastic analysis [Mal], [Nu] (and may be used to produce concentration inequalities on more general path spaces).

The so-called forward and backward martingale method was put forward in the paper [L-Z], and further developed by M. Takeda [Tak]. The exposition here is taken from the notes [Ly] by T. Lyons to which we refer for applications to geometric potential theory.

3. CONCENTRATION AND GEOMETRY

In this chapter, we discuss the concept of Lévy families and its geometric counterparts motivated by the examples presented in the preceding chapter. Following the sphere example, the Lévy families describe asymptotic concentration as the dimension goes to infinity. In the first section, we describe how spectral properties entail exponential concentration, both in continuous and discrete settings. Applications to spectral and diameter bounds are the topic of Section 3.2. In particular, with the tool of concentration, we recover Cheng's upper bound on the diameter of a Riemannian manifold with non-negative Ricci curvature. We then present and study the Lévy families. Topological applications to fixed point theorems emphasized by M. Gromov and V. Milman are further motivation for this geometric investigation. The last application is the historical starting point of the development of the concentration of measure phenomenon. V. Milman [Mi3] indeed used concentration on high dimensional spheres in the early 1970's to produce an inspiring proof of Dvoretzky's theorem about almost spherical sections of convex bodies. We present this proof below with the help of Gaussian rather than spherical concentration as put forward in [Pis2].

3.1 Spectrum and concentration

In this section, we show how spectral properties may yield concentration properties. Assume we are given a smooth compact Riemannian manifold X with Riemannian metric g. Denote by V the total volume of X and by $d\mu = \frac{dv}{V}$ the normalized Riemannian volume element on (X, g). In Chapter 2, we described normal concentration properties of μ under the rather stringent assumption that the Ricci curvatures of (X, g) are strictly positive. In this section, we bound the concentration function $\alpha_{(X,g,\mu)}$ by the spectrum of the Laplace operator Δ on (X, g). More precisely, denote by $\lambda_1 = \lambda_1(X) > 0$ the first non-trivial eigenvalue of Δ. It is well known (cf. [Cha1], [G-H-L]) that λ_1 is characterized by the variational inequality

$$\lambda_1 \operatorname{Var}_\mu(f) \leq \int f(-\Delta f)d\mu = \int |\nabla f|^2 d\mu \tag{3.1}$$

for all smooth real-valued f's on (X, g) where

$$\operatorname{Var}_\mu(f) = \int f^2 d\mu - \left(\int f d\mu\right)^2$$

is the *variance* of f with respect to μ and where $|\nabla f|$ is the Riemannian length of the gradient of f. The historical such inequality (3.1) is the so-called *Poincaré inequality* on the circle (with $\lambda_1 = 1$), which is established by comparing, for a mean zero function f, the L^2-norms of f and f' along the trigonometric basis.

Theorem 3.1. *Let (X,g) be a compact Riemannian manifold with normalized Riemannian measure μ. Then, (X,d,μ) has exponential concentration*

$$\alpha_{(X,g,\mu)}(r) \le \mathrm{e}^{-r\sqrt{\lambda_1}/3}, \quad r > 0,$$

where $\lambda_1 > 0$ is the first non-trivial eigenvalue of the Laplace operator Δ on (X,g).

Proof. Let A and B be (open) subsets of X such that $d(A,B) = \varepsilon > 0$. Set $\mu(A) = a$, $\mu(B) = b$. Consider the function f given by

$$f(x) = \frac{1}{a} - \frac{1}{\varepsilon}\Big(\frac{1}{a} + \frac{1}{b}\Big) \min\big(d(x,A), \varepsilon\big), \quad x \in X$$

(so that $f = 1/a$ on A and $f = 1/b$ on B). By a simple regularization argument, we may apply (3.1) to this f. We need then simply evaluate appropriately the various terms of this inequality. Clearly, f is Lipschitz, $\nabla f = 0$ on $A \cup B$ and

$$|\nabla f| \le \frac{1}{\varepsilon}\Big(\frac{1}{a} + \frac{1}{b}\Big)$$

almost everywhere. Therefore

$$\int |\nabla f|^2 d\mu \le \frac{1}{\varepsilon^2}\Big(\frac{1}{a} + \frac{1}{b}\Big)^2 (1 - a - b).$$

On the other hand,

$$\begin{aligned} \mathrm{Var}_\mu(f) &= \int \big(f - \textstyle\int f d\mu\big)^2 d\mu \\ &\ge \int_A \big(f - \textstyle\int f d\mu\big)^2 d\mu + \int_B \big(f - \textstyle\int f d\mu\big)^2 d\mu \\ &\ge \frac{1}{a} + \frac{1}{b}. \end{aligned}$$

It thus follows from (3.1) that

$$\lambda_1 \varepsilon^2 \le \Big(\frac{1}{a} + \frac{1}{b}\Big)(1 - a - b) \le \frac{1 - a - b}{ab}$$

so that

$$b \le \frac{1-a}{1 + \lambda_1 \varepsilon^2 a} \le \frac{1}{1 + \lambda_1 \varepsilon^2 a}.$$

Choosing for A the complement of B_ε and assuming that $\mu(B_\varepsilon) = 1 - a \le \frac{1}{2}$ shows that

$$\mu(B_\varepsilon) \ge \Big(1 + \frac{\lambda_1 \varepsilon^2}{2}\Big)\mu(B).$$

In other words, the expansion coefficient of μ on (X,g) satisfies

$$\mathrm{Exp}_\mu(\varepsilon) \ge 1 + \frac{\lambda_1 \varepsilon^2}{2} > 1.$$

Set then $\varepsilon > 0$ so that $\lambda_1 \varepsilon^2 = 2$ and the conclusion follows from Proposition 1.13. The numerical constant 3 is not sharp at all.

The proof of Theorem 3.1 is complete. □

On the standard n-sphere $\mathbb{S}^n$, $\lambda_1 = n$ [Cha1] so that Theorem 3.1 yields a weaker concentration than Theorem 2.3. By Lichnerowicz's lower bound $\lambda_1 \geq \frac{nc(X)}{n-1}$ ([G-H-L], [Cha2]) on a compact Riemannian manifold (X, g) with dimension n and strictly positive lower bound $c(X)$ on its Ricci curvature, the same comment applies with respect to Theorem 2.4. However, Theorem 3.1 holds true on any compact manifold.

It is a simple yet non-trivial observation (see Corollary 5.7 below for a proof) that $\lambda_1(X \times Y) = \min(\lambda_1(X), \lambda_1(Y))$ for Riemannian manifolds X and Y. Theorem 3.1 therefore provides a useful tool to concentration in product spaces. In particular, if X^n is the n-fold Riemannian product of a compact Riemannian manifold (X, g) equipped with the product measure μ^n, then

$$\mathrm{ObsDiam}_{\mu^n}(X^n) = O(1).$$

This is in contrast with the Cartesian product for which we observed in (1.25) that $\mathrm{ObsDiam}_{\mu^n}(X^n)$ is essentially of the order of $\sqrt{n}$. This observation is the first step towards dimension free measure concentration for ℓ^2-metrics that will be investigated deeply in Chapter 4 and subsequent.

The proof of Theorem 3.1 generalizes to probability measures μ on a metric space (X, d) that satisfies a Poincaré inequality

$$\mathrm{Var}_\mu(f) \leq C \int |\nabla f|^2 d\mu \tag{3.2}$$

with respect to some (generalized length of) gradient $|\nabla f|$. One may for example consider, given a locally Lipschitz function f on (X, d), the *length of the gradient* of f at the point $x \in X$ to be

$$|\nabla f|(x) = \limsup_{y \to x} \frac{|f(x) - f(y)|}{d(x, y)}.$$

Note that if μ on (X, d) satisfies a Poincaré inequality (3.2) with constant C, the pushed forward measure μ_φ by a 1-Lipschitz map $\varphi : (X, d) \to (Y, \delta)$ also satisfies (3.2) with constant C. Eigenfunction expansions and decompositions along orthogonal polynomials are a fruitful source of examples of measures satisfying (3.2) (see for example [Kl], [K-L-O], etc. and the references therein). For example, the Hermite basis of the L^2-space over the the canonical Gaussian measure γ on $\mathbb{R}^n$ shows that γ satisfies (3.2) with $C = 1$. Rather than spectrum, we actually use Poincaré inequalities towards concentration through the energy $\int |\nabla f|^2 d\mu$. The next statement is an immediate adaptation of Theorem 3.1.

Corollary 3.2. *Assume that μ is a probability measure on the Borel sets of a metric space (X, d) such that for some $C > 0$ and all locally Lipschitz real-valued functions f on (X, d),*

$$\mathrm{Var}_\mu(f) \leq C \int |\nabla f|^2 d\mu.$$

Then

$$\alpha_{(X,g,\mu)}(r) \leq \mathrm{e}^{-r/3\sqrt{C}}, \quad r > 0.$$

Theorem 3.1 has an analogue on graphs to which we turn now. It is convenient to deal with finite state Markov chains.

Let X be a finite or countable set. Let $\Pi(x,y) \geq 0$ satisfy

$$\sum_{y \in X} \Pi(x,y) = 1$$

for every $x \in X$. Assume furthermore that there is a symmetric invariant probability measure μ on X, that is, $\Pi(x,y)\mu(\{x\})$ is symmetric in x and y, and $\sum_x \Pi(x,y)\mu(\{x\}) = \mu(\{y\})$ for every $y \in X$. In other words, (Π,μ) is a reversible *Markov chain.* Define

$$\mathcal{Q}(f,f) = \frac{1}{2} \sum_{x,y \in X} \big[f(x) - f(y)\big]^2 \Pi(x,y)\mu(\{x\}).$$

We may speak of the spectral gap, or rather the Poincaré constant, of the chain (Π,μ) as the largest $\lambda_1 > 0$ such that for all f's,

$$\lambda_1 \mathrm{Var}_\mu(f) \leq \mathcal{Q}(f,f). \tag{3.3}$$

Set also

$$|||f|||_\infty^2 = \frac{1}{2} \sup_{x \in X} \sum_{y \in X} \big|f(x) - f(y)\big|^2 \Pi(x,y).$$

The triple norm $|||\cdot|||_\infty$ may be thought of as a discrete version of the Lipschitz norm in the continuous setting. Although it may not be well adapted to all discrete structures, it behaves similarly for what concerns spectrum and exponential concentration. Theorem 3.3 below is the analogue of Theorem 3.1 in this discrete setting. We however adopt a different strategy of proof best adapted to this case. The argument may be applied similarly to yield alternative proofs of both Theorem 3.1 and Corollary 3.2.

Theorem 3.3. *Let (Π,μ) be a reversible Markov chain on X, as before, with a spectral gap $\lambda_1 > 0$. Then, whenever $|||F|||_\infty \leq 1$, F is integrable with respect to μ and for every $r \geq 0$,*

$$\mu\big(\{F \geq \textstyle\int F d\mu + r\}\big) \leq 3\,\mathrm{e}^{-r\sqrt{\lambda_1}/2}.$$

Proof. Let F be a bounded mean zero function on X with $|||F|||_\infty \leq 1$. Set $\Lambda(\lambda) = \int \mathrm{e}^{\lambda F} d\mu$, $\lambda \geq 0$. We apply the Poincaré inequality to $\mathrm{e}^{\lambda F/2}$. The main observation is that

$$\mathcal{Q}\big(\mathrm{e}^{\lambda F/2}, \mathrm{e}^{\lambda F/2}\big) \leq |||F|||_\infty^2 \int \mathrm{e}^{\lambda F} d\mu. \tag{3.4}$$

Indeed, for $\lambda \geq 0$, by symmetry,

$$\begin{aligned}\mathcal{Q}\big(\mathrm{e}^{\lambda F/2}, \mathrm{e}^{\lambda F/2}\big) &= \frac{1}{2} \sum_{x,y \in X} \big[\mathrm{e}^{\lambda F(x)/2} - \mathrm{e}^{\lambda F(y)/2}\big]^2 \Pi(x,y)\mu(\{x\}) \\ &= \sum_{F(y)<F(x)} \big[\mathrm{e}^{\lambda F(x)/2} - \mathrm{e}^{\lambda F(y)/2}\big]^2 \Pi(x,y)\mu(\{x\}) \\ &\leq \frac{\lambda^2}{2} \sum_{x,y \in X} \big[F(x) - F(y)\big]^2 \mathrm{e}^{\lambda F(x)} \Pi(x,y)\mu(\{x\})\end{aligned}$$

from which (3.4) follows by definition of $|||F|||_\infty$. Therefore,

$$\Lambda(\lambda) - \Lambda\Big(\frac{\lambda}{2}\Big)^2 \leq \frac{\lambda^2}{\lambda_1}\, \Lambda(\lambda),$$

that is, for every $\lambda < \sqrt{\lambda_1}$,

$$\Lambda(\lambda) \leq \frac{1}{1 - \lambda^2/\lambda_1}\, \Lambda\Big(\frac{\lambda}{2}\Big)^2.$$

Applying the same inequality for $\lambda/2$ and iterating, yields, after n steps,

$$\Lambda(\lambda) \leq \prod_{k=0}^{n-1} \Big(\frac{1}{1 - \lambda^2/4^k\lambda_1}\Big)^{2^k} \Lambda\Big(\frac{\lambda}{2^n}\Big)^{2^n}.$$

Since $\Lambda(\lambda) = 1 + o(\lambda)$, we have that $\Lambda(\lambda/2^n)^{2^n} \to 1$ as $n \to 0$. Therefore,

$$\Lambda(\lambda) \leq \prod_{k=0}^{\infty} \Big(\frac{1}{1 - \lambda^2/4^k\lambda_1}\Big)^{2^k}$$

where the product converges whenever $\lambda < \sqrt{\lambda_1}$. The proof of the proposition is easily completed. Indeed, setting for example $\lambda = \frac{1}{2}\sqrt{\lambda_1}$ yields that

$$\Lambda\Big(\frac{\sqrt{\lambda_1}}{2}\Big) = \int \mathrm{e}^{\sqrt{\lambda_1} F/2} d\mu \leq 3.$$

By Chebyshev's inequality

$$\mu\big(\{F \geq r\}\big) \leq 3\,\mathrm{e}^{-r\sqrt{\lambda_1}/2}$$

for every $r \geq 0$. As in Proposition 1.7, the result is easily extended to arbitrary F with $|||F|||_\infty \leq 1$ which completes the proof. □

We may define a distance on X associated with $|||\cdot|||_\infty$ as

$$d(x,y) = \sup_{|||f|||_\infty \leq 1} \big[f(x) - f(y)\big], \quad x,y \in X.$$

Then, together with Proposition 1.7, Theorem 3.3 indicates that

$$\alpha_{(X,d,\mu)}(r) \leq 3\,\mathrm{e}^{-r\sqrt{\lambda_1}/4}, \quad r > 0. \tag{3.5}$$

The distance most often used is however not d but the combinatorial distance d_c associated with the graph with vertex-set X and edge-set $\{(x,y) : \Pi(x,y) > 0\}$. This distance can be defined as the minimum number of edges one has to cross to go from x to y. Equivalently,

$$d_c(x,y) = \sup_{\|\nabla f\|_\infty \leq 1} \big[f(x) - f(y)\big]$$

where

$$\|\nabla f\|_\infty = \sup\big\{|f(x) - f(y)|; \Pi(x,y) > 0\big\}.$$

Now since $\sum_y \Pi(x,y) = 1$,

$$|||f|||_\infty^2 \leq \frac{1}{2}\,\|\nabla f\|_\infty^2.$$

In particular, $d_c \leq d/\sqrt{2}$ and thus (3.5) also holds for d_c.

As an example, let $X = (V, \mathcal{E})$ be a finite connected oriented graph with symmetric set of vertices V and set of edges $\mathcal{E}$. Equip V with the normalized uniform measure and the combinatorial metric d_c. We may consider $\Pi(x,y) = \frac{1}{k(x)}$ whenever x and y are adjacent in V, and 0 otherwise, where $k(x)$ is the number of neighbors of x. Consider the quadratic form

$$\widetilde{\mathcal{Q}}(f,f) = \sum_{x \sim y} \big[f(x) - f(y)\big]^2$$

where the sum runs over all neighbors $x \sim y$ in X. Since X is connected, $\widetilde{\mathcal{Q}}(f,f) \geq 0$ for all f's and is zero whenever f is constant. Let $\lambda_1 > 0$ be the first non-trivial eigenvalue of $\widetilde{\mathcal{Q}}$, that is, of the Laplace operator on X. As a consequence of Theorem 3.3, we have the following result.

Corollary 3.4. *Let $k_0 = \max\{k(x); x \in V\} < \infty$. Then*

$$\alpha_{(X,d_c,\mu)}(r) \leq 3\,\mathrm{e}^{-r\sqrt{\lambda_1/16k_0}}, \quad r > 0.$$

The most important examples of applications of the preceding corollary are the Cayley graphs. Recall that if V is a finite group and $S \subset V$ a symmetric set of generators of V, we may join x and y in V by an edge if $x = s^{-1}y$ for some $s \in S$. The path distance on $X = (V, \mathcal{E})$ is the word distance in V induced by S and $k_0 = \mathrm{Card}(S)$.

Spectral and Poincaré inequalities have an L^1 counterpart related to Cheeger's isoperimetric constant that describes the correspondence between the previous concentration results and the isoperimetric approach of Proposition 2.12. If for example (X, g) is a compact Riemannian manifold with normalized Riemannian measure μ, denote by $h > 0$ the best constant such that

$$h \int \Big|f - \int f\,d\mu\Big|\,d\mu \leq \int |\nabla f|\,d\mu \tag{3.6}$$

for all smooth functions f on (X,g). Cheeger's inequality [Chee] indicates that

$$\lambda_1 \geq \frac{h^2}{4}\,. \tag{3.7}$$

Applying actually (3.6) to characteristic functions of sets amounts to

$$\mu^+(A) \geq 2h\,\mu(A)\big(1-\mu(A)\big),$$

which is close to the isoperimetric inequality (2.13). By Cheeger's inequality (3.7), Theorem 3.1 thus covers the isoperimetric approach in this case, producing similar exponential concentration under weaker geometric invariants. The same comments are more or less in order on graphs (cf. [Al-M], [Alon], [B-H-T]).

3.2 Spectral and diameter bounds

On the basis of the results of Section 3.1, we investigate here some relationships between spectral and diameter bounds.

Assume we are given a smooth complete Riemannian manifold (X,g) that is not necessarily compact but with finite volume V. We denote as usual by $d\mu = \frac{dv}{V}$ the normalized Riemannian volume element. Denote by $\lambda_1 = \lambda_1(X)$ the first non-trivial eigenvalue of the Laplace operator on (X,g).

If $B(x,r)$ is the (open) ball with center x and radius $r>0$ in (X,g), it follows from Theorem 3.1 applied to the 1-Lipschitz function $d(x,\cdot)$ that $\lambda_1 = \lambda_1(X) = 0$ as soon as

$$\limsup_{r\to\infty} \frac{1}{r}\,\log\Big(1-\mu\big(B(x,r)\big)\Big) = 0 \tag{3.8}$$

for some (all) x in X. The following is a kind of converse.

Theorem 3.5. *Let (X,g) be a smooth complete Riemannian manifold with dimension n and finite volume. Let μ be the normalized Riemannian volume on (X,g). Assume that the Ricci curvature of (X,g) is bounded below. Then (X,g) is compact as soon as*

$$\liminf_{r\to\infty} \frac{1}{r}\,\log\Big(1-\mu\big(B(x,r)\big)\Big) = -\infty$$

for some (or all) $x\in X$. In particular $\lambda_1>0$ under this condition. Furthermore, if (X,g) has non-negative Ricci curvature and if D is the diameter of (X,g),

$$D \leq \frac{C_n}{\sqrt{\lambda_1}} \tag{3.9}$$

where $C_n>0$ only depends on the dimension n of X.

The upper bound (3.9) goes back to the work of S.-Y. Cheng [Chen] in Riemannian geometry (see also [Cha1]).

Proof. We proceed by contradiction and assume that X is not compact. Choose $B(x,r_0)$ a geodesic ball in (X,g) with center x and radius $r_0>0$ such that $\mu(B(x,r_0)) \geq \frac{1}{2}$. By non-compactness (and completeness), for every $r>0$, we can take z at distance $r_0 + 2r$ from x. In particular, $B(x, r_0) \subset B(z, 2(r_0 + r))$. By the Riemannian volume comparison theorem [C-E], [Cha2], for every $y \in X$ and $0 < s < t$,

$$\frac{\mu(B(y,t))}{\mu(B(y,s))} \leq \left(\frac{t}{s}\right)^n \mathrm{e}^{t\sqrt{(n-1)K}} \tag{3.10}$$

where $-K$, $K \geq 0$, is a lower bound on the Ricci curvature of (X, g). Therefore,

$$\begin{aligned}\mu\big(B(z,r)\big) &\geq \left(\frac{r}{2(r_0+r)}\right)^n \mathrm{e}^{-2(r+r_0)\sqrt{(n-1)K}}\, \mu\big(B(z, 2(r_0+r))\big) \\ &\geq \frac{1}{2}\left(\frac{r}{2(r_0+r)}\right)^n \mathrm{e}^{-2(r_0+r)\sqrt{(n-1)K}}\end{aligned}$$

where we used that $\mu(B(z, 2(r_0 + r))) \geq \mu(B(x, r_0)) \geq \frac{1}{2}$. Since $B(z, r)$ is included in the complement of $B(x, r_0 + r)$,

$$1 - \mu\big(B(x, r + r_0)\big) \geq \frac{1}{2}\left(\frac{r}{2(r_0+r)}\right)^n \mathrm{e}^{-2(r_0+r)\sqrt{(n-1)K}} \tag{3.11}$$

which is impossible as $r \to \infty$ by the assumption. The first part of the theorem is established.

Thus (X, g) is compact. Denote by D its diameter. Assume that (X, g) has non-negative Ricci curvature. That is, we may take $K = 0$ in (3.10). By Theorem 3.1, for every measurable subset A in X such that $\mu(A) \geq \frac{1}{2}$, and every $r > 0$,

$$1 - \mu(A_r) \leq \mathrm{e}^{-\sqrt{\lambda_1}\, r/3}. \tag{3.12}$$

Let $B(x, \frac{D}{8})$ be the ball with center x and radius $\frac{D}{8}$. We distinguish between two cases. If $\mu(B(x, \frac{D}{8})) \geq \frac{1}{2}$, apply (3.12) to $A = B(x, \frac{D}{8})$. By definition of D, we may choose $r = r_0 = \frac{D}{8}$ in (3.11) to get

$$\frac{1}{2 \cdot 4^n} \leq 1 - \mu(A_{D/8}) \leq \mathrm{e}^{-\sqrt{\lambda_1}\, D/24}.$$

If $\mu(B(x, \frac{D}{8})) < \frac{1}{2}$, apply (3.12) to A, the complement of $B(x, \frac{D}{8})$. Since the ball $B(x, \frac{D}{16})$ is included in the complement of $A_{D/16}$ and since by (3.10) with $t = D$,

$$\mu\Big(B\Big(x, \frac{D}{16}\Big)\Big) \geq \frac{1}{16^n},$$

we get from (3.12) with $r = \frac{D}{16}$ that

$$\frac{1}{16^n} \leq \mathrm{e}^{-\sqrt{\lambda_1}\, D/48}.$$

The conclusion easily follows from either case. Theorem 3.5 is established. □

In the last part of this section, we describe analogous conclusions in the discrete case. As in Section 3.1, let $\Pi(x, y)$ be a Markov chain on a finite state space X with symmetric invariant probability measure μ. Denote by $\lambda_1 > 0$ the spectral gap of (Π, μ) defined by

$$\lambda_1 \mathrm{Var}_\mu(f) \leq \mathcal{Q}(f, f)$$

for every f on X. Recall that here

$$\mathcal{Q}(f,f) = \frac{1}{2}\sum_{x,y\in X}\big[f(x)-f(y)\big]^2\Pi(x,y)\mu(\{x\}).$$

If d is the distance defined from the norm

$$|||f|||_\infty^2 = \frac{1}{2}\sup_{x\in X}\sum_{y\in X}\big|f(x)-f(y)\big|^2\Pi(x,y);$$

recall also from Theorem 3.3 and (3.5) that

$$\alpha_{(X,d,\mu)}(r) \le 3\,\mathrm{e}^{-r\sqrt{\lambda_1}/4}, \quad r>0. \tag{3.13}$$

Denote by D the diameter of X for the distance d.

Proposition 3.6. *If μ is nearly constant, that is, if there exists $C>0$ such that, for every x, $\mu(\{x\}) \le C\min_{y\in X}\mu(\{y\})$, then*

$$\lambda_1 \le \left(\frac{8\log(12C\,\mathrm{Card(X)})}{D}\right)^2.$$

Proof. Consider two points $x,y\in X$ such that $d(x,y)=D$. By Corollary 1.4 and (3.13),

$$\mu(\{x\})\mu(\{y\}) \le 12\,\mathrm{e}^{-D\sqrt{\lambda_1}/8}.$$

Since, by the hypothesis on μ, $\min_{z\in X}\mu(\{z\}) \ge (C\,\mathrm{Card}(X))^{-1}$, the conclusion follows. □

Recall that the combinatorial diameter D_c is such that $D_c \le D/\sqrt{2}$.

3.3 Lévy families

In this section, we deal with families of metric measure spaces (X^n,d^n) equipped with Borel probability measures μ^n, $n\ge 1$. Denote by D^n the diameter of (X^n,d^n) and assume that $1\le D^n<\infty$, $n\ge 1$. According to the definition of the concentration function, we say that the family $(X^n,d^n,\mu^n)_{n\ge 1}$ is a *Lévy family* if for every $r>0$,

$$\lim_{n\to\infty}\alpha_{(X^n,d^n,\mu^n)}(D^n r) = 0. \tag{3.14}$$

More quantitatively, we say that $(X^n,d^n,\mu^n)_{n\ge 1}$ is a *normal Lévy family* if there are constants $C,c>0$ such that for every $n\ge 1$ (or only n large enough) and $r>0$,

$$\alpha_{(X^n,d^n,\mu^n)}(r) \le C\,\mathrm{e}^{-cnr^2}. \tag{3.15}$$

Since we assumed that $D^n\ge 1$, any normal Lévy family is a Lévy family. The omission of D^n in the definition of a normal Lévy family is justified by the fact that many examples become that way Lévy families with their natural metrics.

A particular situation occurs when μ^n, $n \geq 1$, is a family of probability measures on a given fixed metric space (X,d). We then say similarly that $(X,d,\mu^n)_{n\geq 1}$ is a Lévy family if for every $r > 0$,

$$\lim_{n\to\infty} \alpha_{(X,d,\mu^n)}(r) = 0. \tag{3.16}$$

The study of Chapter 2 provides a number of geometric examples of (normal) Lévy families. Let us review a few of them. The family $(\mathbb{S}^n_{R_n}, \sigma^n_{R_n})$ of the Euclidean spheres with radii R_n equipped with normalized Haar measures is a Lévy family as soon as $n^{-1/2}R_n \to 0$ as $n \to \infty$. It is a normal Lévy family if $R_n \sim 1$. More generally, by Theorem 2.4, a family $(X^n, g^n)_{n\geq 1}$ of compact Riemannian manifolds equipped with the normalized volume elements μ^n such that $c(X^n) \to \infty$ with n is a Lévy family. Recall that we denote by $c(X)$ the infimum of the Ricci curvature tensor over all unit tangent vectors of a Riemannian manifold (X,g). This includes $\mathbb{S}^n$, $\mathbb{SO}^n$ for which $c(\mathbb{SO}^n) = \frac{n-1}{4}$, the Stiefeld manifolds $\mathbb{W}^n_k$, etc. (cf. [Mi1], [C-E], [Cha2]). By Propositions 2.8 and 2.9, we may add the examples of unit cubes $[0,1]^n$ (with normalized Euclidean metric) and Euclidean unit balls which form normal Lévy families for their uniform measures.

If $(X^n, g^n)_{n\geq 1}$ are compact Riemannian manifolds such that $\lambda_1(X^n) \to \infty$ as $n \to \infty$ where $\lambda_1(X^n)$ is the first non-trivial eigenvalue of the Laplace operator on X^n, then $(X^n, g^n)_{n\geq 1}$ is a Lévy family by Theorem 3.1, however usually not normal.

The discrete cubes $\{0,1\}^n$ with normalized Hamming metrics and counting measures define a normal Lévy family. There are many known families of k_0-regular graphs X (with k_0 fixed) such that $\mathrm{Card}(X) \to \infty$ whereas $\lambda_1 \geq \epsilon > 0$ stays bounded away from zero (the so-called expanders graphs). Moreover graphs with this property are "generic" amongst k_0-regular graphs [Alo]. These examples thus give also rise to Lévy families. Further combinatorial examples such as the symmetric group Π^n will be described in the next chapter.

The following simple proposition is an alternative definition of Lévy families. Recall that if A is a subset of a metric space (X,d) and if $r > 0$, we let $A_r = \{x \in X; d(x,A) < r\}$.

Proposition 3.7. *Let $(X^n, d^n, \mu^n)_{n\geq 1}$ be a Lévy family. Then, for any sequence $A^n \subset X^n$, $n \geq 1$, such that $\liminf_{n\to\infty} \mu^n(A^n) > 0$, we have for every $\varepsilon > 0$,*

$$\lim_{n\to\infty} \mu^n\big((A^n)_{\varepsilon D^n}\big) = 1.$$

When $(X^n, d^n, \mu^n)_{n\geq 1}$ is a normal Lévy family, the same result holds with εD^n replaced by ε.

Proof. Let $\liminf_{n\to\infty} \mu^n(A^n) > \eta > 0$. Since $(X^n, d^n, \mu^n)_{n\geq 1}$ is a Lévy family, for every $r > 0$, $\alpha_{\mu^n}(D^n r) < \eta$ for all n's large enough. Therefore by Lemma 1.1, for every n large enough and every $s > 0$,

$$1 - \mu^n\big((A^n)_{D^n(r+s)}\big) \leq \alpha_{\mu^n}(D^n s).$$

Using again that $\alpha_{\mu^n}(D^n s) \to 0$ as $n \to \infty$, and since $r, s > 0$ are arbitrary, the conclusion follows. □

The next statement describes the general properties of Lévy families that immediately follow from the definition and results of Chapter 1.

Proposition 3.8. *Let $(X^n, d^n, \mu^n)_{n\geq 1}$ be a (normal) Lévy family.*

For every $n \geq 1$, let φ^n be a Lipschitz map between (X^n, d^n) and (Y^n, δ^n). Denote by $\mu^n_{\varphi^n}$ the image measure of μ^n by φ^n, $n \geq 1$. If $\sup_{n\geq 1} \|\varphi^n\|_{\mathrm{Lip}} < \infty$, then $(Y^n, \delta^n, \mu^n_{\varphi^n})_{n\geq 1}$ is again a (normal) Lévy family.

If $(Y^n, \delta^n, \mu^n)_{n\geq 1}$ is another (normal) Lévy family, the product family $(X^n \times Y^n, d^n + \delta^n, (\mu^n \otimes \nu^n)_{n\geq 1})_{n\geq 1}$ is a (normal) Lévy family.

Let $A^n \subset X^n$, $n \geq 1$, Borel sets such that $\liminf_{n\to\infty} \mu^n(A^n) > 0$. Then $(X^n, d^n, \mu^n(\cdot\,|A_n))_{n\geq 1}$ is a (normal) Lévy family.

3.4 Topological applications

In this section, we present applications of the concentration of measure phenomenon to some fixed point theorems.

Let (X, d) be a metric space, and let G be a family of maps from X into X. We say that a subset A of X is *essential* (with respect to the action of G) if for every $\varepsilon > 0$ and every finite subset $\{g_1, \ldots, g_\ell\} \subset G$,

$$\bigcap_{k=1}^{\ell} g_k(A_\varepsilon) \neq \emptyset.$$

(Recall $A_\varepsilon = \{x \in X; d(x, A) < \varepsilon\}$. We then say that the pair (X, G) has the *property of concentration* if for every finite covering $X \subset \bigcup_{i=1}^N A^i$, $A^i \subset X$, there exists A^i which is essential (for the action of G).

The following proposition connects this definition with Lévy families.

Proposition 3.9. *Let (X, d) and G be as before. Suppose that $G = \bigcup_{n\geq 1} G^n$, $G^n \subset G^{n+1}$, and that there exist probability measures μ^n, $n \geq 1$, on the Borel sets of (X, d) such that μ^n is G^n-invariant for each n. If $(X, d, \mu^n)_{n\geq 1}$ is a Lévy family, then (X, G) has the property of concentration.*

Proof. Assume $X = \bigcup_{i=1}^N A^i$. There exists i, $1 \leq i \leq N$, such that

$$\limsup_{n\to\infty} \mu^n(A^i) \geq \frac{1}{2N}\,.$$

Selecting a subsequence n', by Proposition 3.7, for every $\varepsilon > 0$,

$$\lim_{n'\to\infty} \mu^{n'}\big((A^i)_\varepsilon\big) = 1.$$

Since μ^n is G^n-invariant, for every $g \in G$,

$$\lim_{n'\to\infty} \mu^{n'}\big(g((A^i)_\varepsilon)\big) = 1.$$

It immediately follows that A^i is essential. □

The following examples illustrate the proposition. Let H be an infinite dimensional Hilbert space and let $(e_i)_{i\geq 1}$ be an orthonormal basis of H. The orthogonal group $\mathbb{SO}^n$ may be realized as unitary operators on H which are the identity on the span of $(e_i)_{i>n}$. Then $\mathbb{SO}^n \subset \mathbb{SO}^{n+1}$. Consider then $X = G = \mathbb{SO}^\infty = \bigcup_{n\geq 1} \mathbb{SO}^n$ equipped with the Hilbert-Schmidt operator metric. Then (X, G) has the concentration property. To prove it, recall that $\mathbb{SO}^n$ equipped with the normalized Haar measures μ^n, $n \geq 1$, form a normal Lévy family and apply Proposition 3.9 (with μ^n defined on all of X).

Exactly the same proof together with concentration on spheres shows that the unit sphere S^∞ of H with the action of $G = \mathbb{SO}^\infty$ has the concentration property. One may also show that if u is a unitary operator on H and if $G = (u^n)_{n\geq 1}$, then (S^∞, G) has the property of concentration in the above sense.

If G consists of a family of pairwise commuting unitary operators on H, then (S^∞, G) has still the property of concentration. However, this is no more true if G is the group of all unitary operators. Recently, V. Pestov [Pe1], [Pe2] described conditions on G in terms of amenability for the property of concentration to hold. Let π be a unitary representation of a group G in H. In particular, G acts on the unit sphere S_H of the space of representation.

Theorem 3.10. *A discrete group is amenable if and only if the dynamical system (S_H, G, π) has the property of concentration for every unitary representation π of G in H.*

If G' is a subgroup of G and (S_H, G) has the property of concentration, then (S_H, G') also has the property of concentration. For example, if F_2 denotes the free group on two generators, the pair formed by the unit sphere of the space $\mathrm{L}^2(F_2)$ and the full unitary group of this space does not have the property of concentration.

Moreover, amenable representations can be characterized in terms of the concentration property. The proof of Theorem 3.10 is based on proper extensions of fixed point theorems under the property of concentration to which we turn now. Actually, the concentration property of the system (S_H, G, π) has to do not with the amenability of the acting group G as such, but rather with the amenability of the representation π. We refer to [Pe1] for the proof of Theorem 3.10 and for extensions to locally compact groups as well as for dynamical and ergodic applications.

As announced, we turn to fixed point theorems under the property of concentration. Let (X, d) be a metric space. Let G be a family of continuous maps from X into X and suppose the identity map belongs to G. We call the pair (X, G) a *G-space*. We do not assume that G is a group (or even a semigroup) although this will mostly be the case.

Proposition 3.11. *Let (X, d) be a G-space and let Γ be a compact space. Assume that (X, G) has the property of concentration. Let $\psi : X \to \Gamma$ be any map with domain X. Then there exists $x \in \Gamma$ such that for any neighborhood O of x, the set $\psi^{-1}(O)$ is essential in X.*

We call such point x an essential point of the map ψ.

Proof. We proceed by contradiction. If such x does not exist, we can find for every y in Γ an open neighborhood O_y such that $\psi^{-1}(O_y)$ is not essential. Since Γ is compact, there is a finite family $(O_i)_{1\leq i\leq N}$ of the covering $(O_y)_{y\in\Gamma}$ of Γ such that

$\Gamma \subset \bigcup_{i=1}^{N} O_i$. Therefore $\bigcup_{i=1}^{N} \psi^{-1}(O_i) = X$ and, by the concentration property, one of the sets $\psi^{-1}(O_i)$ has to be essential which is a contradiction. The proposition is proved. □

A map $\psi : X \to \Gamma$ where Γ is compact is called uniformly continuous if for any closed subset $A \subset \Gamma$ and any open neighborhood O_A of A, there is $\varepsilon > 0$ such that

$$\psi\Big(\big[\psi^{-1}(A)\big]_{\varepsilon} \Big) \subset O_A.$$

Theorem 3.12. *Let (X, d) be a G-space and let Γ be a compact space. Assume that (X, G) has the property of concentration. Let $\psi : X \to \Gamma$ be uniformly continuous and $g \in G$ be fixed, and assume that there exists a continuous $J : \Gamma \to \Gamma$ such that $Jg = \psi g$. Then any essential point of ψ is a fixed point for J.*

Before turning to the proof of Theorem 3.12, let us mention the following immediate consequence that motivated the present investigation and that follows by choosing $\Gamma = X$, ψ the identity map and $J = g$.

Corollary 3.13. *Let (X, d) be a compact G-space. If (X, G) has the property of concentration, then there is a point $x \in X$ which is fixed under the action of G.*

More generally, if $\psi : X \to \Gamma$ is *equivariant* in the sense that for every $g \in G$ there exists a continuous $J_g : \Gamma \to \Gamma$ such that $J_g \psi = \psi g$, then any essential point of ψ is a fixed point for J_g, $g \in G$. One example is provided by a (compact) group G acting on Γ, fixing $x \in \Gamma$, and setting $\psi : G \to \Gamma$, $\psi g = gx$.

Proof of Theorem 3.12. Let x be an essential point of ψ such that $J(x) = y \neq x$. By continuity of J, there exist open neighborhoods O_x of x and O_y of y such that $J(O_x) \subset O_y$ and $O_x \cap O_y = \emptyset$. We may moreover choose O_x such that for some neighborhoods U and V of the closures of O_x and O_y, we have $U \cap V = \emptyset$. Since x is an essential point of ψ, $\psi^{-1}(O_x) = \chi \subset X$ is an essential set. Therefore, for every $\varepsilon > 0$, $(g(\chi))_\varepsilon \cap \chi_\varepsilon \neq \emptyset$. Since ψ is uniformly continuous, there exists $\varepsilon > 0$ such that $\psi(\chi)_\varepsilon \subset U$ and $\psi([\psi^{-1}(O_y)]_\varepsilon) \subset V$. Since

$$g(\chi) \subset \psi^{-1}\big(J(O_x)\big) \quad \text{and} \quad J(O_x) \subset O_y,$$

we get $\psi((g(\chi))_\varepsilon) \subset V$. Now, $(g(\chi))_\varepsilon \cap \chi_\varepsilon \neq \emptyset$, so that $U \cap V \neq \emptyset$ which yields a contradiction. The proof of Theorem 3.12 is complete. □

In the final part of this section, we turn to some finite dimensional analogue for which we may not claim fixed points but for which we estimate the diameter of minimal invariant subspaces.

Let (X, d) be a compact metric space with a Borel probability measure μ. We denote by α_μ the concentration function of (X, d, μ). Also let G be a family of measure-preserving maps from X into X.

Let (Γ, δ) be a compact metric space and let $\psi : X \to \Gamma$ be an equivariant map. We assume that $\|J_g\|_{\mathrm{Lip}} \leq L$ for every J_g. Denote by $N_\Gamma(\varepsilon)$ the minimum number of (open) balls of radius $\varepsilon > 0$ which cover Γ.

Theorem 3.14. *Under the above conditions and notations, there exists $x \in \Gamma$ such that for every J_g, $g \in G$,*

$$d(J_g x, x) \leq (1 + L) \inf \big(\varepsilon + \|\psi\|_{\mathrm{Lip}}(s + r)\big)$$

where the infimum runs over all $\varepsilon, s, r > 0$ such that $N_\Gamma(\varepsilon)\alpha_\mu(s) < 1$ and $\alpha_\mu(r) < \frac{1}{2}$.

Proof. Let $\varepsilon > 0$ and $s > 0$ satisfy the required condition and choose a finite subset $\mathcal{S}$ of Γ with cardinality $N_\Gamma(\varepsilon)$ such that the balls centered in $\mathcal{S}$ with radius ε cover Γ. Choose $x \in \mathcal{S}$ such that

$$\mu\big(\psi^{-1}(B(x,\varepsilon))\big) \geq \frac{1}{N_\Gamma(\varepsilon)} > \alpha_\mu(s)$$

where $B(x,\varepsilon)$ is the ball with center x and radius ε in Γ. Define $\chi = \psi^{-1}(B(x,\varepsilon))$. By Lemma 1.1,

$$\mu(\chi_{s+r}) \geq 1 - \alpha_\mu(r) > \frac{1}{2}$$

when $\alpha_\mu(r) < \frac{1}{2}$. Now,

$$\psi(\chi_{s+r}) \subset B\big(x, \varepsilon + (s+r)\|\psi\|_{\mathrm{Lip}}\big) = B_1$$

while, by the definition of L, for any $g \in G$,

$$J_g(B_1) \subset B\Big(J_g x, L\big(\varepsilon + (s+r)\|\psi\|_{\mathrm{Lip}}\big)\Big) = B_2.$$

Therefore, $\mu(\psi^{-1}(B_1)) \geq \mu(\chi_{s+r}) \geq \frac{1}{2}$ while

$$\mu\big(\psi^{-1}(B_2)\big) \geq \mu\big(\psi^{-1}J_g(B_1)\big) = \mu\big(g\psi^{-1}(B_1)\big) > \frac{1}{2}\,.$$

It follows that $B_1 \cap B_2 \neq \emptyset$ so that

$$d(J_g x, x) \leq (1+L)\big(\varepsilon + \|\psi\|_{\mathrm{Lip}}(s+r)\big).$$

Since g is arbitrary in G, the theorem is established. □

A typical application occurs when Γ is a compact set in $\mathbb{R}^k$ equipped with a norm $\|\cdot\|$ in which case $N_\Gamma(\varepsilon) \leq (1 + \frac{2R}{\varepsilon})^k$ where R is the radius of Γ for the norm $\|\cdot\|$ (see Lemma 3.18 below). Then, provided that μ has normal concentration $\alpha_\mu(r) \leq C\,\mathrm{e}^{-cnr^2}$, $r > 0$, it may be shown from Theorem 3.14 that $R \leq \mathrm{Const}\sqrt{\frac{k}{n}}$ where the constant depends on the various parameters in Theorem 3.14 (see [Mi4] for details).

3.5 Euclidean sections of convex bodies

This last section is devoted to the historical example of application of concentration to the geometric problem of spherical sections of convex bodies.

Let $\mathcal{K}$ be a convex symmetric body in $\mathbb{R}^n$, that is, a convex compact subset of $\mathbb{R}^n$ with non-empty interior, symmetric with respect to the origin. $\mathcal{K}$ is said to contain almost Euclidean sections of dimension k if, for every $\varepsilon > 0$, there exist a k-dimensional subspace H and an ellipsoid $\mathcal{E}$ in H such that

$$(1-\varepsilon)\,\mathcal{E} \subset \mathcal{K} \cap H \subset (1+\varepsilon)\,\mathcal{E}. \tag{3.17}$$

This geometric description admits the following equivalent functional formulation. A Banach space E (with norm $\|\cdot\|$) is said to contain a subspace $(1+\varepsilon)$-isomorphic to the Euclidean space $\mathbb{R}^k$ if there are vectors $v_1, \dots, v_k$ in E such that for all $t = (t_1, \dots, t_k)$ in $\mathbb{R}^k$,

$$(1-\varepsilon)\,|t| \le \Big\| \sum_{i=1}^{k} t_i v_i \Big\| \le (1+\varepsilon)\,|t| \tag{3.18}$$

where $|\cdot|$ is the Euclidean norm on $\mathbb{R}^k$. Indeed, if $\|\cdot\|$ is the gauge of the convex body $\mathcal{K}$, the subspace H generated by $v_1, \dots, v_k$ and the ellipsoid $\mathcal{E}$, the image of the Euclidean unit ball of $\mathbb{R}^k$ under the isomorphism $e_i \to v_i$, satisfy (3.17).

The important next statement is the famous Dvoretzky theorem in the local theory of normed spaces.

Theorem 3.15. *For each $\varepsilon > 0$ there exists $\eta(\varepsilon) > 0$ such that every Banach space E of dimension n contains a subspace $(1+\varepsilon)$-isomorphic to $\mathbb{R}^k$ where $k = [\eta(\varepsilon)\log n]$.*

Since every $2n$-dimensional ellipsoid has an n-dimensional section which is a multiple of the canonical Euclidean ball, we see from Theorem 3.15 that for each $\varepsilon > 0$ there exists $\eta(\varepsilon) > 0$ such that every centrally symmetric convex body $\mathcal{K}$ admits a central section $\mathcal{K}_0$ with dimension $k \ge \eta(\varepsilon)\log n$ and $\rho > 0$ such that $(1-\varepsilon)\rho\mathcal{B}^k \subset \mathcal{K}_0 \subset (1+\varepsilon)\rho\mathcal{B}^k$ where $\mathcal{B}^k$ is the canonical Euclidean ball in the subspace spanned by $\mathcal{K}_0$.

The rest of the section is devoted to the proof of this result. While the original concentration proof by V. Milman [Mi3] uses high dimensional spherical concentration, we rather make use of Gaussian concentration following [Pis2] (see also [Pis3]). The main argument will be to span the Euclidean subspace by the rotational invariance of Gaussian distributions. The first lemma, known as the Dvoretzky-Rodgers lemma, will be crucial in the choice of the underlying Gaussian distribution.

Lemma 3.16. *Let $(E, \|\cdot\|)$ be a Banach space of dimension n and let $m = [n/2]$ (integer part). There exist vectors $w_1, \dots, w_m$ in E such that $\|w_j\| \ge 1/2$ for all $j = 1, \dots, m$ and*

$$\Big\| \sum_{j=1}^{m} t_j w_j \Big\| \le |t|$$

for all $t \in \mathbb{R}^m$.

Proof. We first construct an operator $T : \mathbb{R}^n \to E$ such that $\|T\| \le 1$ and $\|T_{|V}\| \ge \dim(V)/n$ for all subspaces V of $\mathbb{R}^n$. In particular $\|T\| = 1$. To this task, consider any determinant function (associated to any fixed matrix representation) $S \to \det(S)$. Let T be such that

$$\det(T) = \max\big\{\det(S); S : \mathbb{R}^n \to E, \|S\| \le 1\big\}.$$

Then, for any $\varepsilon > 0$ and any $S : \mathbb{R}^n \to E$,

$$\det(T + \varepsilon S) \le \det(T)\|T + \varepsilon S\|^n. \tag{3.19}$$

Now,

$$\begin{aligned}\det(T + \varepsilon S) &= \det(T)\det(\mathrm{Id} + \varepsilon T^{-1}S)\\ &= \det(T)\big(1 + \varepsilon\,\mathrm{Tr}(T^{-1}S) + o(\varepsilon)\big).\end{aligned}$$

Hence, substituting in (3.19), we get

$$\begin{aligned}1+\varepsilon\,\mathrm{Tr}(T^{-1}S) &\le \|T+\varepsilon S\|^n + o(\varepsilon)\\ &\le \big(1+\varepsilon\|S\|\big)^n + o(\varepsilon)\\ &\le 1+\varepsilon n\|S\| + o(\varepsilon)\end{aligned}$$

so that $\mathrm{Tr}(T^{-1}S) \le n\|S\|$. Consider now a subspace V in $\mathbb{R}^n$ and let $P : \mathbb{R}^n \to V$ be the orthogonal projection. Setting $S = TP$ we get

$$\dim(V) = \mathrm{Tr}(P) = \mathrm{Tr}(T^{-1}S) \le n\|S\|.$$

Therefore $\|TP\| \ge \dim(V)/n$ which proves the claim.

It is now easy to complete the proof of the lemma. Let T be as before. Choose y_1 in $\mathbb{R}^n$ such that $|y_1| = 1$ and $\|Ty_1\| = \|T\| = 1$. Then choose y_2 orthogonal to y_1 such that $|y_2| = 1$ and $\|Ty_2\| = \|TQ\| \ge (1-\frac{1}{n})$ where Q is the orthogonal projection onto $\{y_1\}^\perp$. Continuing in this way on the basis of the preceding claim, one constructs a sequence such that $y_j \in \{y_1, \dots, y_{j-1}\}^\perp$ and $|y_j| = 1$, $\|Ty_j\| \ge (n-j+1)/n$. Set then $w_j = Ty_j$, $j = 1, \dots, [n/2]$, which have the required properties. The proof of Lemma 3.16 is complete. □

Define the map $F : \mathbb{R}^m \to \mathbb{R}$ by

$$F(x) = \Big\|\sum_{j=1}^m x_j w_j\Big\|, \quad x = (x_1, \dots, x_m) \in \mathbb{R}^m. \tag{3.20}$$

By Lemma 3.16, $\|F\|_{\mathrm{Lip}} \le 1$. Indeed, for all $x, y \in \mathbb{R}^m$,

$$\big|F(x) - F(y)\big| \le \Big\|\sum_{j=1}^m (x_j - y_j)w_j\Big\| \le |x-y|.$$

Denoting by γ the canonical Gaussian measure on $\mathbb{R}^m$, by (2.35) for example, for every $r \ge 0$,

$$\gamma\big(\{|F - \textstyle\int F d\gamma| \ge r\}\big) \le 2\,\mathrm{e}^{-r^2/2}. \tag{3.21}$$

It is an important feature of the construction of Lemma 3.16 that $M_F = \int F d\gamma$ is much greater (for n large) than $\|F\|_{\mathrm{Lip}} \le 1$. This is the effect of high dimensional geometry in this Gaussian framework. Namely, denoting by μ the uniform measure on the discrete cube $\{-1,+1\}^m$, by symmetry of γ,

$$M_F = \iint \Big\|\sum_{j=1}^m \varepsilon_j x_j w_j\Big\| d\mu(\varepsilon) d\gamma(x).$$

By Jensen's inequality, conditionally on ε_k, $k \ne j$, it follows that

$$M_F \ge \int \max_{1\le j\le m} \|x_j w_j\| d\gamma(x) \ge \frac{1}{2}\int \max_{1\le j\le m} |x_j| d\gamma(x)$$

where we used Lemma 3.16 in the second step. By a well-known and elementary computation, there is a numerical constant $\rho > 0$ such that

$$\int \max_{1\le j\le m} |x_j| d\gamma(x) \ge 2\rho\sqrt{\log m}\,.$$

Hence, for every $n \geq 3$,

$$M_F \geq \rho \sqrt{\log n}\,. \tag{3.22}$$

As a next step, we need two technical lemmas to discretize the problem of finding Euclidean subspaces. A δ-net ($\delta > 0$) of $A \subset \mathbb{R}^k$ is a subset $\mathcal{S} \subset A$ such that for every t in A one can find s in $\mathcal{S}$ with $|t - s| \leq \delta$.

Lemma 3.17. *For each $\varepsilon > 0$ there is $\delta = \delta(\varepsilon) > 0$ with the following property. Let $\mathcal{S}$ be a δ-net of the unit sphere of $\mathbb{R}^k$. Then, if for some $v_1, \dots, v_k$ in a Banach space E, we have $1 - \delta \leq \|\sum_{i=1}^k t_i v_i\| \leq 1 + \delta$ for all t in $\mathcal{S}$, then*

$$(1-\varepsilon)\,|t| \leq \Big\|\sum_{i=1}^k t_i v_i\Big\| \leq (1+\varepsilon)\,|t|$$

for all t in $\mathbb{R}^k$.

Proof. It is enough, by homogeneity, to establish the conclusion for every t in $\mathbb{R}^k$ with $|t| = 1$. By definition of $\mathcal{S}$, there exists t^0 in $\mathcal{S}$ such that $|t - t^0| \leq \delta$, hence $t = t^0 + \lambda_1 s$ with s in the unit sphere of $\mathbb{R}^k$ and $|\lambda_1| \leq \delta$. Iterating the procedure, we get that $t = \sum_{\ell=0}^\infty \lambda_\ell t^\ell$ with $t^\ell \in \mathcal{S}$ and $|\lambda_\ell| \leq \delta^\ell$ for all $\ell \geq 0$. Hence

$$\Big\|\sum_{i=1}^k t_i v_i\Big\| \leq \sum_{\ell=0}^\infty \delta^\ell \Big\|\sum_{i=1}^k t_i^\ell v_i\Big\| \leq \frac{1+\delta}{1-\delta}$$

($\delta < 1$). In the same way,

$$\Big\|\sum_{i=1}^k t_i v_i\Big\| \geq 1 - 2\delta\,.$$

It therefore suffices to choose δ appropriately, as a function of $\varepsilon > 0$ only, in order to get the conclusion. □

The size of a δ-net of spheres in finite dimension is easily estimated in terms of $\delta > 0$ and the dimension, as is shown by the next lemma which follows from a simple volume argument.

Lemma 3.18. *There is a δ-net $\mathcal{S}$ of the unit sphere of $\mathbb{R}^k$ of cardinality less than $(1 + \frac{2}{\delta})^k \leq e^{2k/\delta}$.*

Proof. Let $t^1, \dots, t^\ell$ be maximal in the unit sphere of $\mathbb{R}^k$ under the relations $|t^i - t^j| > \delta$ for $i \neq j$. Then the balls in $\mathbb{R}^k$ with centers t^i and radius $\frac{\delta}{2}$ are disjoint and are all contained in the ball with center at the origin and radius $1 + \frac{\delta}{2}$. By comparing the volumes, we get $\ell(\frac{\delta}{2})^k \leq (1 + \frac{\delta}{2})^k$ from which the lemma follows. □

We are now in a position to prove Theorem 3.15. The main argument is the concentration property for Gaussian measures (3.21). The idea is to span the Euclidean subspace by the Gaussian rotational invariance. Consider the Lipschitz function F defined in (3.20) and recall the canonical Gaussian measure γ on $\mathbb{R}^m$, $m = [n/2]$. Let $k \geq 1$ be specified later on and let $t = (t_1, \dots, t_k)$ in the unit sphere of $\mathbb{R}^k$ (i.e., $|t| = 1$). By the Gaussian rotational invariance, the distribution of the map

$$(X_1, \dots, X_k) \in (\mathbb{R}^m)^k \mapsto \sum_{i=1}^k t_i X_i \in \mathbb{R}^m$$

under the product measure γ^k on $(\mathbb{R}^m)^k$ is the same as γ. Therefore, by (3.21), for every $r \geq 0$,

$$\gamma^k\left(\left\{(X_1,\ldots,X_k)\in(\mathbb{R}^m)^k;\left|F\left(\sum_{i=1}^k t_iX_i\right)-M_F\right|\geq r\right\}\right)\leq 2\,\mathrm{e}^{-r^2/2}.$$

Now let $\varepsilon > 0$ be fixed and choose $\delta = \delta(\varepsilon) > 0$ according to Lemma 3.17. Furthermore, let $\mathcal{S}$ be a δ-net in the unit sphere of $\mathbb{R}^k$, which can be chosen with cardinality less than or equal to $\mathrm{e}^{2k/\delta}$ (Lemma 3.18). Let $r = \delta M_F$. The preceding inequality together with the definition of F thus implies that

$$\begin{aligned}\gamma^k\left(\left\{\exists t\in\mathcal{S};\left|\left\|\sum_{i=1}^k t_i\frac{Z_i}{M_F}\right\|-1\right|\geq\delta\right\}\right)&\leq 2\,\mathrm{Card}\,(\mathcal{S})\,\mathrm{e}^{-\delta^2M_F^2/2}\\&\leq 2\,\mathrm{e}^{(2k/\delta)-(\delta^2M_F^2/2)}\end{aligned}$$

where

$$Z_i=\sum_{j=1}^m X_{ij}w_j\in E,\quad X_i=(X_{ij})_{1\leq j\leq m},\ i=1,\ldots,k.$$

Assume n to be large enough (otherwise there is nothing to prove). Since $M_F \geq \rho\sqrt{\log n}$, we can then choose $k = [\eta\log n]$ for $\eta = \eta(\delta) = \eta(\varepsilon)$ small enough such that the preceding probability is strictly less than 1. It follows that there is at least one realization of $(X_1,\ldots,X_k)$ such that $(v_1,\ldots,v_k) = (M_F)^{-1}(Z_1,\ldots,Z_k)$ satisfy

$$1-\delta\leq\left\|\sum_{i=1}^k t_iv_i\right\|\leq 1+\delta$$

for all t in S. Together with Lemma 3.17, this completes the proof of the theorem. □

The preceding proof of Theorem 3.15 actually shows that most random sections (in the sense of γ^k) will produce almost Euclidean subspaces. Further use of concentration arguments (by means of empirical processes) shows that the dependence of ε of the function $\eta(\varepsilon)$ is of the order of ε^2 [Sche2]. The order $k \sim \log n$ is optimal as shown by the example of the cube. ℓ^p-balls in $\mathbb{R}^n$ have Euclidean sections proportional to $n^{2/p}$ for $p \geq 2$ and of the order of n for $1 \leq p \leq 2$ [F-L-M].

As announced, the original measure concentration argument by V. Milman [Mi3] used spherical rather than Gaussian concentration. One nice feature of the spherical proof is the following geometric consequence [Mi3], [Mi5].

Theorem 3.19. *For each $\eta > 0$ there exists $\delta = \delta(\eta) > 0$ such that for any continuous function F on $\mathbb{S}^n$, there exists a k-dimensional sphere $\mathbb{S}^k \subset \mathbb{S}^n$ (i.e., the unit sphere of a $(k+1)$-dimensional subspace) with $k = [\delta n]$ such that for any x in $\mathbb{S}^k$,*

$$\left|F(x)-m_F\right|<\omega_F(2\eta)$$

where m_F is a median of F (for σ^n) and ω_F is the modulus of continuity of F.

Proof. It is a consequence of Lévy's inequality (2.6). Consider the set

$$A=\{|F-m_F|<\omega_F(\eta)\}.$$

By (2.6), we know that

$$\sigma^n(A) \geq 1 - 2\,\mathrm{e}^{(n-1)\eta^2/2}.$$

Fix x in $\mathbb{S}^n$ and let μ be the Haar probability measure on $\mathbb{SO}^n$. Clearly,

$$\mu\big(\{T \in \mathbb{SO}^n; Tx \in A\}\big) = \sigma^n(A) \geq 1 - 2\,\mathrm{e}^{-(n-1)\eta^2/2}.$$

This implies that for any finite set $\mathcal{S}$ in $\mathbb{S}^n$,

$$\mu\big(\{T \in \mathbb{SO}^n; T\mathcal{S} \subset A\}\big) \geq 1 - 2\,\mathrm{Card}(\mathcal{S})\,\mathrm{e}^{-(n-1)\eta^2/2}.$$

Therefore, if $2\,\mathrm{Card}(\mathcal{S}) < \mathrm{e}^{(n-1)\eta^2/2}$, we may find a rotation T such that $T\mathcal{S}$ is a subset of A. Choose then $\mathcal{S}$ to be a δ-net of $\mathbb{S}^k$ according to Lemma 3.18. This is thus possible for $k = [\delta n]$ for some appropriate $\delta = \delta(\eta) > 0$ from which the result follows. □

Notes and Remarks

Dvoretzky's theorem on Euclidean subspaces of Banach spaces (Theorem 3.15) was established first in [Dv]. The new proof by V. Milman [Mi3] was at the starting point of the developments of the concentration ideas. Since then, V. Milman indeed strongly promoted the usefulness of the concentration of measure phenomenon in various contexts [Mi5], and convinced in particular M. Talagrand of the importance of this simple, but useful, property. The proof presented in Section 3.5 of Dvoretzky's theorem is the Gaussian version, due to G. Pisier [Pis2], [Pis3], of Milman's argument (that used concentration on high dimensional spheres). Lemma 3.16 is an important step known as the Dvoretzky-Rodgers lemma [D-R]. Applications of concentration to Dvoretzky-like theorems have been a main issue in the local theory of Banach spaces in the decades 1970-90, starting with the fundamental contribution [F-L-M] by T. Figiel, J. Lindenstrauss and V. Milman where the connection with type and cotype of Banach spaces is amply demonstrated. The main results are extensively reviewed in the Lecture Notes [M-S] by V. Milman and G. Schechtman and more recently in the Handbook in the geometry of Banach spaces [J-L] (cf. the contributions [Sche5], [Gi-M1], [J-S3], etc.) See also [Pis3]. In particular, fine embeddings of subspaces of L^p in finite dimensional ℓ^p-spaces and finite dimensional ℓ^p-subspaces, $1 < p \leq 2$, in connection with stable type, have been investigated by W. Johnson and G. Schechtman [J-S1] and G. Pisier [Pis1] together with further probabilistic concentration inequalities for series of independent random vectors (cf. [Pis2], [Sche5]). Approximation of zonoids by zonotopes by random embeddings in [B-L-M] makes use of concentration through empirical process methods as initiated in [Sche3]. Theorem 3.19 is taken from Milman's original paper [Mi3]. V. Milman also used concentration for some infinite dimensional integration questions and other problems [Mi2].

Concentration under spectral properties was first put forward by M. Gromov and V. Milman in [Gr-M1] from which Theorem 3.1 and its proof are taken. A proof of Theorem 3.1 using a variation on the Herbst argument of Chapter 5 is presented in [Schmu2]. The corresponding results on graphs and applications to expander graphs and superconcentrators have been investigated by N. Alon and V. Milman [Al-M],

[Alo]. See also [B-H-T]. Further geometric examples may be found in [Grom2]. The proof of Theorem 3.3 is taken from [A-S]. Connections between concentration and Poincaré's inequalities strongly developed on the probabilistic side during the past decade (cf. [Le5] and the references therein).

The results of Section 3.2 are inspired by early contributions of R. Brooks [Bro]. Theorem 3.5 is implicit in [Le5] where the analogous results between bounds on the diameter and the logarithmic Sobolev constant are developed. Cheng's inequality (3.9) was established in [Chen].

The Lévy families are introduced and analyzed from a geometric and topological point of view by M. Gromov and V. Milman in [Gr-M1] where a number of geometric examples are provided. The results of Section 3.3 and the fixed point theorems of Section 3.4 are taken from this reference. See also [Grom2], [Mi5]. In particular, the interested reader will find in [Grom2] inspiring developments on convergence of metric measure spaces related to the Lévy families. Theorem 3.14 is due to V. Milman [Mi4]. Theorem 3.10 is one example of the recent deep investigation by V. Pestov [Pe1], [Pe2] to connect measure concentration and ergodic theory by showing that an action on some group has concentration if and only if it is amenable.

4. CONCENTRATION IN PRODUCT SPACES

While products of Lévy families are still Lévy families, at a more quantitative level the concentration functions of product spaces are very sensible to dimension. An important issue addressed in this chapter is to describe dimension free concentration properties in product spaces. With respect to the high dimensional effects of concentration developed in Chapter 3, the normalization chosen here points towards the dimension free Gaussian model. Several methods are developed to this task, and further ones will be presented in Chapters 5 and 6.

We already described in Chapter 1 concentration in product spaces for the ℓ^1-metric and the Hamming metric. In this chapter (and the subsequent ones), we rather deal with the ℓ^2-metric to produce dimension free concentration properties. We start with a useful martingale method that discretizes some of the martingale tools of Chapter 2. We next present some aspects of the deep investigation by M. Talagrand [Tal7] of concentration in product spaces. Following his work, we analyze successively the convex hull and q-point approximations of powerful use in discrete and combinatorial probability theory as illustrated in Section 8.4. In particular, while the basic definition of concentration deals with metric spaces, we investigate here examples in product spaces that go far outside the usual metric setting. Proofs rely on a basic induction scheme together with geometric arguments. Section 4.4 is devoted to the application of infimum-convolution inequalities to concentration for product measures as already outlined at the end of the first chapter. The last part describes the striking concentration phenomenon for the exponential distribution that goes beyond the concentration phenomenon for Gaussian measures. From a probabilistic point of view, most of the concentration and deviation inequalities we will obtain furthermore extend classical inequalities on sums of independent random variables to arbitrary Lipschitz (or Lipschitz and convex) functionals of the sample, leading to powerful tools in applications (see Chapters 7 and 8).

4.1 Martingale methods

We examine first some useful martingale inequalities that yield a variety of further concentration results.

Let $(\Omega, \mathcal{F}, \mathbb{P})$ be a probability space. Denote by $\mathbb{E}(f)$ the expectation of an integrable real-valued function f on $(\Omega, \mathcal{F})$ with respect to $\mathbb{P}$. Consider a finite *filtration* of sub-σ-fields

$$\{\emptyset, \Omega\} = \mathcal{F}_0 \subset \mathcal{F}_1 \subset \cdots \subset \mathcal{F}_n = \mathcal{F}$$

of the σ-field $\mathcal{F}$. For every $i = 1, \dots, n$, set

$$d_i = \mathbb{E}^{\mathcal{F}_i}(f) - \mathbb{E}^{\mathcal{F}_{i-1}}(f)$$

where $\mathbb{E}^{\mathcal{F}_i}$ denotes conditional expectation with respect to the sub-σ-field $\mathcal{F}_i$. The first lemma bounds the deviation of a function f from its mean in terms of the sizes of the increments d_i. It is the discrete analogue of (2.37).

Lemma 4.1. *Let f be integrable with respect to $\mathbb{P}$. For every $r \geq 0$,*

$$\mathbb{P}\big(\{f \geq \mathbb{E}(f) + r\}\big) \leq \mathrm{e}^{-r^2/2D^2}$$

where $D^2 \geq \sum_{i=1}^n \|d_i\|_\infty^2$.

Applied simultaneously to $-f$, Lemma 4.1 yields the concentration of f around its mean,

$$\mathbb{P}\big(\{|f - \mathbb{E}(f)| \geq r\}\big) \leq 2\,\mathrm{e}^{-r^2/2D^2}. \tag{4.1}$$

Proof. By convexity, for every $\lambda \in \mathbb{R}$ and $-1 \leq u \leq +1$,

$$\mathrm{e}^{\lambda u} \leq \frac{1+u}{2}\,\mathrm{e}^{\lambda} + \frac{1-u}{2}\,\mathrm{e}^{-\lambda}.$$

Hence, by homogeneity and since $\mathbb{E}^{\mathcal{F}_{i-1}}(d_i) = 0$, for every $\lambda \in \mathbb{R}$,

$$\mathbb{E}^{\mathcal{F}_{i-1}}(\mathrm{e}^{\lambda d_i}) \leq \cosh\big(\lambda \|d_i\|_\infty\big) \leq \mathrm{e}^{\lambda^2 \|d_i\|_\infty^2/2}.$$

The properties of conditional expectations then prove that

$$\mathbb{E}\big(\mathrm{e}^{\sum_{i=1}^n d_i}\big) = \mathbb{E}\big(\mathrm{e}^{\sum_{i=1}^{n-1} d_i}\mathbb{E}^{\mathcal{F}_{n-1}}(\mathrm{e}^{\lambda d_n})\big) \leq \mathrm{e}^{\lambda^2 \|d_n\|_\infty^2/2}\mathbb{E}\big(\mathrm{e}^{\sum_{i=1}^{n-1} d_i}\big).$$

Iterating,

$$\mathbb{E}\big(\mathrm{e}^{\sum_{i=1}^n d_i}\big) \leq \mathrm{e}^{\lambda^2 D^2/2}.$$

Since $f - \mathbb{E}(f) = \sum_{i=1}^n d_i$, by Chebyshev's exponential inequality, for every $\lambda, r \geq 0$,

$$\mathbb{P}\big(\{f \geq \mathbb{E}(f) + r\}\big) \leq \mathrm{e}^{-\lambda r}\,\mathbb{E}\big(\mathrm{e}^{\sum_{i=1}^n d_i}\big) \leq \mathrm{e}^{-\lambda r + \lambda^2 D^2/2}.$$

Minimizing in $\lambda \geq 0$ thus yields the inequality of the lemma. □

Lemma 4.1 is the martingale extension of the Hoeffding type inequality (1.23). Inequalities of this type have been proved extremely useful in the study of limit theorems in classical probability theory and in discrete algorithmic mathematics [MD2].

Lemma 4.1 is of special interest once the decomposition

$$\{\emptyset, \Omega\} = \mathcal{F}_0 \subset \mathcal{F}_1 \subset \cdots \subset \mathcal{F}_n = \mathcal{F}$$

in sub-σ-fields is such that the martingale differences d_i of a given function f can be controlled. The following metric decomposition is a useful tool in this regard.

A finite metric space (X, d) is of *length* at most ℓ if there exist an increasing sequence $\{X\} = \mathcal{X}^0, \mathcal{X}^1, \dots, \mathcal{X}^n = \{\{x\}\}_{x \in X}$ of partitions of X ($\mathcal{X}^i$ is a refinement of $\mathcal{X}^{i-1}$) and positive numbers $a_1, \ldots, a_n$ with $\ell = \left(\sum_{i=1}^n a_i^2\right)^{1/2}$ such that, if $\mathcal{X}^i = \{A_j^i\}_{1 \leq j \leq m_i}$, for all $i = 1, \ldots, n$, $p = 1, \ldots, m_{i-1}$ and j, k such that $A_j^i, A_k^i \subset A_p^{i-1}$, there exists a one-to-one and onto function $\phi : A_j^i \to A_k^i$ such that $d(x, \phi(x)) \leq a_i$ for every $x \in A_j^i$. Note that ℓ is always smaller than or equal to the diameter of X.

Theorem 4.2. *Let (X, d) be a finite metric space of length at most ℓ, and let μ be the normalized counting measure on X. Then, for every 1-Lipschitz function F on (X, d) and every $r \geq 0$,*

$$\mu\big(\{F \geq \textstyle\int F d\mu + r\}\big) \leq \mathrm{e}^{-r^2/2\ell^2}.$$

In particular,

$$\alpha_{(X,d,\mu)}(r) \leq \mathrm{e}^{-r^2/8\ell^2}, \quad r > 0.$$

Before turning to the proof of Theorem 4.2, let us present examples that motivate the definition of spaces with controlled length in the above sense. First, let X be the discrete cube $\{0,1\}^n$ equipped with the normalized Hamming metric and normalized counting (product) measure $\mu = \mu^n$. The obvious choice of partitions (adding one coordinate at each step) shows that $\{0,1\}^n$ is of length at most $\frac{1}{\sqrt{n}}$. Therefore, we deduce from Theorem 4.2 that

$$\alpha_{(\{0,1\}^n,\mu)}(r) \leq \mathrm{e}^{-nr^2/8}, \quad r > 0. \tag{4.2}$$

Compare with Theorem 2.11.

Actually, the same would apply to any product measure with respect to the Hamming metric. Indeed, let $(\Omega_i, \Sigma_i, \mu_i)$, $i = 1, \ldots, n$, be arbitrary finite probability spaces and let P be the product measure $\mu_1 \otimes \cdots \otimes \mu_n$ on the product space $X = \Omega_1 \times \cdots \times \Omega_n$. A point x in X has coordinates $x = (x_1, \ldots, x_n)$. Equip $X = \Omega_1 \times \cdots \times \Omega_n$ with the Hamming metric

$$d(x, y) = \mathrm{Card}\,\{1 \leq i \leq n; x_i \neq y_i\}.$$

Choose the partitions $\mathcal{X}^i$ consisting of the elements

$$\Omega_1 \times \cdots \times \Omega_i \times (x_{i+1}, \ldots, x_n), \quad x_{i+1}, \ldots, x_n \in \Omega.$$

It is clear that the sequence of partitions $(\mathcal{X}^i)_{1 \leq i \leq n}$ fulfils the definition of the length of a metric space with all the a_i's equal to 1 by definition of the Hamming metric. We thus recover in this way (1.24).

A further, non-product, example is given by the symmetric group Π^n of permutations of $\{1, \ldots, n\}$ equipped with the normalized metric

$$d(\sigma, \pi) = \frac{1}{n}\,\mathrm{Card}\,\{i; \sigma(i) \neq \pi(i)\}$$

and uniform measure μ (assigning mass $(n!)^{-1}$ to each permutation). (The choice of the normalized metric is again guided here for the matter of comparison with spherical concentration.) For every $i = 1, \ldots, n$, consider the elements

$$A_{j_1,\ldots,j_i} = \big\{\sigma \in \Pi^n; \sigma(1) = j_1, \ldots, \sigma(i) = j_i\big\}$$

where the $j_1, \dots, j_i$ are distinct in $\{1, \dots, n\}$. These elements form a partition $\mathcal{X}^i$ of Π^n. Now, given $A = A_{j_1,\dots,j_i}$ as above, let $B = A_{j_1,\dots,j_i,p}$ and $C = A_{j_1,\dots,j_i,q}$ be contained in A. Let τ be the transposition that changes p with q. Define $\phi : B \to C$ by letting $\phi(\sigma) = \tau \circ \sigma$. It is easily seen that $d(\sigma, \phi(\sigma)) \le \frac{2}{n}$ so that we may take the a_i's equal to $\frac{2}{n}$.

Corollary 4.3. *Let μ be a uniform measure on the symmetric group Π^n. For any 1-Lipschitz function F on (Π^n, d) and any $r \ge 0$,*

$$\mu\big(\{F \ge \textstyle\int F d\mu + r\}\big) \le e^{-nr^2/8}.$$

In particular,

$$\alpha_{(\Pi^n, d, \mu)}(r) \le e^{-nr^2/32}, \quad r > 0.$$

We now present the proof of Theorem 4.2.

Proof of Theorem 4.2. Denote by $\mathcal{F}_i$ the σ-field generated by the partition $\mathcal{X}^i$, $i = 1, \dots, n$. Given F integrable with respect to P, set $F_i = E^{\mathcal{F}_i}(F)$, $i = 0, 1, \dots, n$ $(F_0 = \mathbb{E}(F))$. For any $B = A^i_j$ and $C = A^i_k$ contained in A^{i-1}_p, F_i is constant on B and C and

$$\big|F_{i|B} - F_{i|C}\big| \le a_i \tag{4.3}$$

where $F_{i|B}$, $F_{i|C}$ denote the restriction of F_i to B, respectively C. Indeed, $F_{i|B} = (\mathrm{Card}\,(B))^{-1} \sum_{\sigma \in B} F(\sigma)$ while

$$F_{i|C} = \big(\mathrm{Card}\,(C)\big)^{-1} \sum_{\sigma \in C} F(\sigma) = \big(\mathrm{Card}\,(B)\big)^{-1} \sum_{\sigma \in B} F\big(\phi(\sigma)\big)$$

so that (4.3) immediately follows.

Therefore, for any A, B as before, $\big|F_{i|B} - F_{i-1|A}\big| \le a_i$. Indeed,

$$F_{i-1|A} = \frac{1}{N} \sum_{C \subset A} F_{i|A}$$

with $N = \mathrm{Card}\,\{C; C \subset A\}$ so that

$$\big|F_{i|B} - F_{i-1|A}\big| \le \frac{1}{N} \sum_{C \subset A} \big|F_{i|B} - F_{i|C}\big| \le a_i.$$

Hence $\|d_i\|_\infty \le a_i$ for any $i = 1, \dots, n$ and Lemma 4.1 yields the desired conclusion. The proof of Theorem 4.2 is complete. □

The examples of the discrete cube and symmetric group suggest one further extension of the method. Given a compact metric group G with a translation invariant metric d, and a closed subgroup H of G, one can define a natural metric δ on the quotient G/H by

$$\delta(gH, g'H) = d(g, g'H) = d\big({g'}^{-1}g, H\big).$$

The translation invariance of d implies that this is actually a metric and that $d(g, g'H)$ does not depend on the representative g of gH.

Theorem 4.4. *Let G be a compact metric group with a translation invariant metric d and let $G = G_0 \supset G_1 \supset \cdots \supset G_n = \{e\}$ be a decreasing sequence of closed subspaces of G. Denote by a_i the diameter of G_{i-1}/G_i, $i = 1, \ldots, n$, and let $\ell = \left(\sum_{i=1}^n a_i^2\right)^{1/2}$. Let μ be Haar measure on G. Then,*

$$\alpha_{(G,d,\mu)}(r) \leq \mathrm{e}^{-r^2/8\ell^2}, \quad r > 0.$$

Proof. We show that for every mean zero 1-Lipschitz function F on (G, d) and every $r \geq 0$,

$$\mu(\{F \geq r\}) \leq \mathrm{e}^{-r^2/2\ell^2}.$$

Let $\mathcal{F}_i$, $i = 0, \ldots, n$, be the σ-algebra generated by the sets $\{gG_i\}_{g \in G}$. Note that if $gG_{i-1} \supset hG_i$, then $g^{-1}h \in G_{i-1}$. If both $gG_{i-1} \supset h_1G_i$ and $gG_{i-1} \supset h_2G_i$, let $s \in G_{i-1}$ be such that

$$\mathrm{Diam}\,(G_{i-1}/G_i) = d(g^{-1}h_1, g^{-1}h_2G_i) = d(g^{-1}h_1, g^{-1}h_2s).$$

Define then $\phi : h_1G_i \to h_2G_i$ by $\phi(h_1r) = h_2sr$. Then

$$d\big(h_1r, \phi(h_1r)\big) = d(h_1, h_2s) = \mathrm{Diam}\,(G_{i-1}/G_i) = a_i.$$

Therefore, if $F_i = \mathbb{E}^{\mathcal{F}_i}(F)$, $i = 0, \ldots, n$, the oscillation of F_i on each atom of $\mathcal{F}_{i-1}$ is at most a_i, so that, in the notation of Lemma 4.1, $\|d_i\|_\infty \leq a_i$. Lemma 4.1 then allows us to complete the proof of the theorem. □

In the case of the symmetric group over n elements, we may take $G = \Pi^n$, $G_i = \Pi^{n-i}$ and $a_i = \frac{2}{n}$. One further instance of this theorem is the n-dimensional torus $\mathbb{T}^n$ equipped with the normalized product measure μ and the normalized ℓ^1-metric $\frac{1}{n}\sum_{i=1}^n |s_i - t_i|$. Taking $\mathbb{T}^i$, $i = 0, \ldots, n$, as subgroups, one gets $a_i \leq \frac{1}{n}$ for every i so that, by Theorem 4.4,

$$\alpha_{(\mathbb{T}^n,d,\mu)}(r) \leq 2\,\mathrm{e}^{-nr^2/8}, \quad r > 0.$$

This result is to be compared with Proposition 2.7 where the stronger Euclidean metric was inforced, for which dimension free concentration holds.

We complete this section by another application of the martingale method to norms of sums of independent random vectors (one can also use the language of supremum of empirical processes as in Chapter 7). Let $Y_1, \ldots, Y_n$ be independent integrable random variables on some probability space $(\Omega, \mathcal{F}, \mathbb{P})$ taking values in a Banach space $(E, \|\cdot\|)$ and let $S = \sum_{i=1}^n Y_i$. Now

$$\|S\| - \mathbb{E}\big(\|S\|\big) = \sum_{i=1}^n d_i$$

can be written as a sum of martingale differences with respect to the filtration $\mathcal{F}_i$ generated by $Y_1, \ldots, Y_i$, $i = 1, \ldots, n$, with the property that, for every $i = 1, \ldots, n$,

$$\begin{aligned}
|d_i| &= \Big|\big(\mathbb{E}^{\mathcal{F}_i} - \mathbb{E}^{\mathcal{F}_{i-1}}\big)\big(\|S\|\big)\Big| \\
&= \Big|\big(\mathbb{E}^{\mathcal{F}_i} - \mathbb{E}^{\mathcal{F}_{i-1}}\big)\big(\|S\| - \|S - Y_i\|\big)\Big| \\
&\leq \|Y_i\| + \mathbb{E}\big(\|Y_i\|\big)
\end{aligned}$$

by the triangle inequality and independence. In a sense, $\|S\| - \mathbb{E}(\|S\|)$ is as good as the sum $\sum_{i=1}^n \|Y_i\|$, so that, provided $\mathbb{E}(\|S\|)$ is under control, the classical one-dimensional results apply. As a consequence of Lemma 4.1, we thus obtain the following extension of (1.23).

Corollary 4.5. *Let $Y_1, \ldots, Y_n$ be independent bounded random vectors in a Banach space $(E, \|\cdot\|)$ and let $S = \sum_{i=1}^n Y_i$. For every $r \geq 0$,*

$$\mathbb{P}\Big(\Big\{\big|\|S\| - \mathbb{E}(\|S\|)\big| \geq r\Big\}\Big) \leq 2\,\mathrm{e}^{-r^2/2D^2}$$

where $D^2 \geq \sum_{i=1}^n \|Y_i\|_\infty^2$.

4.2 Convex hull approximation

In this section, we investigate concentration for product measures through geometric considerations involving convexity properties. Let $(\Omega_i, \Sigma_i, \mu_i)$, $i = 1, \ldots, n$, be arbitrary probability spaces and let P be the product measure $\mu_1 \otimes \cdots \otimes \mu_n$ on $X = \Omega_1 \times \cdots \times \Omega_n$. A point x in X has coordinates $x = (x_1, \ldots, x_n)$.

Recall the Hamming distance on X defined by

$$d(x, y) = \mathrm{Card}\,\{1 \leq i \leq n;\ x_i \neq y_i\}.$$

As we have seen in (1.24) and Section 4.1 above, the concentration function of any product probability P on (X, d) satisfies

$$\alpha_{(X,d,P)}(r) \leq \mathrm{e}^{-cr^2/n}, \quad r > 0, \tag{4.4}$$

for some numerical constant $c > 0$. In particular, if $P(A) \geq \frac{1}{2}$, for most of the elements x in Ω^n, there exists $y \in A$ within distance $\sqrt{n}$ of x.

The same result actually applies to all the Hamming metrics

$$d_a(x, y) = \sum_{i=1}^n a_i \mathbf{1}_{\{x_i \neq y_i\}}, \quad a = (a_1, \ldots, a_n) \in \mathbb{R}_+^n,$$

with $|a|^2 = \sum_{i=1}^n a_i^2$ instead of n on the right-hand side of (4.4). This type of normalization leads to dimension free formulations.

Note that various measurability questions may arise in the subsequent statements. These are actually completely unessential and will be ignored below (start for example with finite or Polish probability spaces Ω_i).

We now introduce a control that will be uniform in the Hamming metrics d_a whenever $|a| \leq 1$. We define, for every non-empty subset A of $X = \Omega_1 \times \cdots \times \Omega_n$ and every $x \in X$, a distance $\mathcal{D}_A^c(x)$ from x to A as

$$\mathcal{D}_A^c(x) = \sup_{|a|=1} d_a(x, A).$$

This definition somewhat hides the combinatorial and convexity properties of the functional $\mathcal{D}_A^c$ that will be needed in its investigation. For a subset $A \subset X$ and $x \in X$, let

$$U_A(x) = \big\{s = (s_i)_{1 \leq i \leq n} \in \{0,1\}^n;\ \exists\, y \in A \text{ such that } y_i = x_i \text{ if } s_i = 0\big\}.$$

One can use equivalently the collection of the characteristic functions $\mathbf{1}_{\{x_i \neq y_i\}}$, $y \in A$. Denote by $V_A(x)$ the convex hull of $U_A(x)$ as a subset of $[0,1]^n \subset \mathbb{R}^n$. Note that $0 \in V_A(x)$ if and only if $x \in A$. One may then measure the distance from x to A by the Euclidean distance $d(0, V_A(x))$ from 0 to $V_A(x)$. It is easily seen that

$$\mathcal{D}_A^c(x) = d\big(0, V_A(x)\big) = \inf_{y \in V_A(x)} |y|. \tag{4.5}$$

We recall that $|\cdot|$ denotes the Euclidean norm on $\mathbb{R}^n$. Indeed, if $d(0, V_A(x)) \le r$, there exists z in $V_A(x)$ with $|z| \le r$. Let $a \in \mathbb{R}_+^n$ with $|a| = 1$. Then

$$\inf_{y \in V_A(x)} (a \cdot y) \le a \cdot z \le |z| \le r$$

where $u \cdot v$ is the scalar product of $u, v \in \mathbb{R}^n$. Since

$$\inf_{y \in V_A(x)} (a \cdot y) = \inf_{s \in U_A(x)} (a \cdot s) = d_a(x, A), \tag{4.6}$$

$\mathcal{D}_A^c(x) \le r$. Conversely, let $z \in V_A(x)$ be such that $|z| = d(0, V_A(x))$ (> 0) and set $a = z/|z|$. Let $y \in V_A(x)$. By convexity, for every $\theta \in [0,1]$, $\theta y + (1-\theta)z \in V_A(x)$ so that

$$\big|z + \theta(y - z)\big|^2 = \big|\theta y + (1-\theta)z\big|^2 \ge |z|^2.$$

Hence, as $\theta \to 0$, $(y - z) \cdot z \ge 0$, that is,

$$a \cdot y \ge |z| = d(0, V_A(x))$$

for every $y \in V_A(x)$. Now, by (4.6),

$$\mathcal{D}_A^c(x) \ge d_a(x, A) = \inf_{y \in V_A(x)} (a \cdot y) \ge d\big(0, V_A(x)\big)$$

which is the claim.

The next theorem extends the concentration (4.4) to this uniformity with dimension free bounds.

Theorem 4.6. *For every measurable non-empty subset A of $X = \Omega^1 \times \cdots \times \Omega^n$, and every product probability measure P on X,*

$$\int \mathrm{e}^{(\mathcal{D}_A^c)^2/4} dP \le \frac{1}{P(A)}.$$

In particular, for every $r \ge 0$,

$$P\big(\{\mathcal{D}_A^c \ge r\}\big) \le \frac{1}{P(A)}\, \mathrm{e}^{-r^2/4}.$$

The general scheme of proof is by induction on the number of coordinates together with geometric argument involving projections and sections to lower dimensional subspaces. The main difficulty in this type of statements is to find the adapted recurrence hypothesis expressed here by the exponential integral inequalities. For simplicity in the notation, we assume that we are given a probability space (Ω, Σ, μ) and that P is the n-fold product μ^n of μ on $X = \Omega^n$. This is no loss of generality. Since the crucial inequalities will not depend on n, we need simply to work on products of $(\widetilde{\Omega}, \widetilde{\mu}) = \big(\Pi_{i=1}^n \Omega_i, \bigotimes_{i=1}^n \mu_i\big)$ with itself and consider the coordinate map

$$\widetilde{x} = (\widetilde{x}_1, \ldots, \widetilde{x}_n) \in \widetilde{\Omega}^n \to (\widetilde{x}_i)_i \in \Omega_i, \quad \widetilde{x}_i = ((\widetilde{x}_i)_1, \ldots, (\widetilde{x}_i)_n) \in \widetilde{\Omega},$$

that only depends on the i-th factor.

Proof of Theorem 4.6. The case $n = 1$ is easy. To go from n to $n+1$, let A be a subset of Ω^{n+1} and let B be its projection on Ω^n. Furthermore, for $\omega \in \Omega$, let $A(\omega)$ be the section of A along ω. If $x \in \Omega^n$ and $\omega \in \Omega$, set $z = (x, \omega)$. The key observation is the following: if $s \in U_{A(\omega)}(x)$, then $(s, 0) \in U_A(z)$, and if $t \in U_B(x)$, then $(t, 1) \in U_A(z)$. It follows that if $\xi \in V_{A(\omega)}(x)$, $\zeta \in V_B(x)$ and $0 \le \theta \le 1$, then $(\theta\xi + (1-\theta)\zeta, 1-\theta) \in V_A(z)$. By the description (4.5) of $\mathcal{D}_A^c$ and convexity of the square function,

$$\begin{aligned}\mathcal{D}_A^c(z)^2 &\le (1-\theta)^2 + \big|\theta\xi + (1-\theta)\zeta\big|^2 \\ &\le (1-\theta)^2 + \theta|\xi|^2 + (1-\theta)|\zeta|^2.\end{aligned}$$

Hence,

$$\mathcal{D}_A^c(z)^2 \le (1-\theta)^2 + \theta \mathcal{D}_{A(\omega)}^c(x)^2 + (1-\theta)\mathcal{D}_B^c(x)^2.$$

Now, by Hölder's inequality and the induction hypothesis, for every ω in Ω,

$$\begin{aligned}&\int_{\Omega^n} \mathrm{e}^{(\mathcal{D}_A^c(x,\omega))^2/4} dP(x) \\ &\qquad \le \mathrm{e}^{(1-\theta)^2/4} \Big(\int_{\Omega^n} \mathrm{e}^{(\mathcal{D}_{A(\omega)}^c)^2/4} dP\Big)^{\theta} \Big(\int_{\Omega^n} \mathrm{e}^{(\mathcal{D}_B^c)^2/4} dP\Big)^{1-\theta} \\ &\qquad \le \mathrm{e}^{(1-\theta)^2/4} \Big(\frac{1}{P(A(\omega))}\Big)^{\theta} \Big(\frac{1}{P(B)}\Big)^{1-\theta},\end{aligned}$$

that is,

$$\int_{\Omega^n} \mathrm{e}^{(\mathcal{D}_A^c(x,\omega))^2/4} dP(x) \le \frac{1}{P(B)}\, \mathrm{e}^{(1-\theta)^2/4} \Big(\frac{P(A(\omega))}{P(B)}\Big)^{-\theta}.$$

Optimize now in θ. To this aim, use that, for every $u \in [0, 1]$,

$$\inf_{\theta \in [0,1]} \mathrm{e}^{(1-\theta)^2/4} u^{-\theta} \le 2 - u. \tag{4.7}$$

Taking $\theta = 1 + 2\log u$ if $u \ge \mathrm{e}^{-1/2}$ and $\theta = 0$ otherwise, and taking logarithms, it suffices to show that

$$\log(2-u) + \log u + (\log u)^2 \ge 0,$$

which is established through elementary calculus. Therefore, by (4.7),

$$\int_{\Omega^n} \mathrm{e}^{(\mathcal{D}_A^c(x,\omega))^2/4} dP(x) \le \frac{1}{P(B)} \Big(2 - \frac{P(A(\omega))}{P(B)}\Big).$$

To conclude, integrate in ω so that, by Fubini's theorem,

$$\begin{aligned}\int_{\Omega^{n+1}} \mathrm{e}^{(\mathcal{D}_A^c(x,\omega))^2/4} dP(x)d\mu(\omega) &\leq \frac{1}{P(B)}\left(2 - \frac{P\otimes\mu(A)}{P(B)}\right)\\ &\leq \frac{1}{P\otimes\mu(A)}\end{aligned}$$

since $u(2-u) \leq 1$ for every real number u. Theorem 4.6 is established. □

The strength of the functional $\mathcal{D}_A^c(x)$ with respect to the Hamming metric is that all choices of a_i's (depending upon x) are possible. This makes Theorem 4.6 a principle of considerable power in applications, as will be further demonstrated in Chapters 7 and 8. As a first illustration, let $\Omega = \{0,1\}$ and μ_p be the probability measure that gives mass p to 1. Denote by $P = \mu_p^n$ the product measure of μ_p on Ω^n. Consider a subset A of Ω^n that is hereditary in the sense that if $x = (x_1,\dots,x_n) \in A$ and if $y = (y_1,\dots,y_n) \in \Omega^n$ is such that $y_i \leq x_i$ for all $i = 1,\dots,n$, then $y \in A$. For $x \in \Omega^n$, set $J = \{1 \leq i \leq n; x_i = 1\}$ and $N(x) = \mathrm{Card}(J)$. Choosing $a_i = 1$ if $i \in J$ and 0 otherwise shows that

$$d(x,A) \leq \mathcal{D}_A^c(x)\sqrt{N(x)}$$

where $d(x,A)$ is the usual Hamming distance from x to A. Therefore, for every $r,s > 0$,

$$\begin{aligned}P\big(\{d(\cdot,A) \geq r\}\big) &\leq P\big(\{\mathcal{D}_A^c \geq rs^{-1/2}\}\big) + P\big(\{y : N(y) > s\}\big)\\ &\leq \frac{1}{P(A)}\,\mathrm{e}^{-r^2/4s} + P\big(\{y : N(y) > s\}\big).\end{aligned}$$

Since the last term is very small for $s > pn$, we see that Theorem 4.6 produces the correct order $(pn)^{-1}$ in the coefficient of r^2.

By definition of the convex hull distance $\mathcal{D}_A^c(x)$ as a supremum of distances, its application to Lipschitz functions requires some care. Consider indeed a function $F : X \to \mathbb{R}$ on the product space $X = \Omega_1 \times \cdots \times \Omega_n$ such that for every $x \in X$ there exists $a = a(x) \in \mathbb{R}_+^n$ with $|a| = 1$ such that for every $y \in X$,

$$F(x) \leq F(y) + d_a(x,y). \tag{4.8}$$

The important feature in the extension provided by Theorem 4.6 is that a may depend on x. Then let $A = \{F \leq m\}$. By (4.8) and the definition of $\mathcal{D}_A^c(x)$, it is clear that

$$F(x) \leq m + d_a(x,A) \leq m + \mathcal{D}_A^c(x)$$

for every x. Hence, by Theorem 4.6, for every $r \geq 0$,

$$P\big(\{F \geq m + r\}\big) \leq P\big(\{\mathcal{D}_A^c \geq r\}\big) \leq \frac{1}{P(A)}\,\mathrm{e}^{-r^2/4}.$$

In other words,

$$P\big(\{F \leq m\}\big)P\big(\{F \geq m + r\}\big) \leq \mathrm{e}^{-r^2/4}, \quad r \geq 0.$$

Choose then successively $m = m_F$ and $m = m_F - r$ where m_F is a median of F for P to get the following result.

Corollary 4.7. *Let P be a product probability measure on the product space $X = \Omega_1 \times \cdots \times \Omega_n$ and let $F : X \to \mathbb{R}$ be 1-Lipschitz in the sense of (4.8). Then, for every $r \geq 0$,*

$$P(\{|F - m_F| \geq r\}) \leq 4\,\mathrm{e}^{-r^2/4}$$

where m_F is a median of F for P.

Replacing F by $-F$, Corollary 4.7 applies similarly if (4.8) is replaced by $F(y) \leq F(x) + d_a(x, y)$.

A typical application of Corollary 4.7 concerns supremum of linear functionals. In a probabilistic language, consider independent real-valued random variables $Y_1, \ldots, Y_n$ on some probability space $(\Omega, \mathcal{A}, \mathbb{P})$ such that for real numbers u_i, v_i, $i = 1, \ldots, n$,

$$u_i \leq Y_i \leq v_i, \quad i = 1, \ldots, n.$$

Set

$$Z = \sup_{t \in \mathcal{T}} \sum_{i=1}^{n} t_i Y_i \tag{4.9}$$

where $\mathcal{T}$ is a (finite or countable) family of vectors $t = (t_1, \ldots, t_n)$ in $\mathbb{R}^n$ such that $\sigma = \sup_{t \in \mathcal{T}} \left(\sum_{i=1}^n t_i^2 (v_i - u_i)^2\right)^{1/2} < \infty$. Apply Corollary 4.7 to

$$F(x) = \sup_{t \in \mathcal{T}} \sum_{i=1}^{n} t_i x_i$$

on $X = \prod_{i=1}^n [u_i, v_i]$ under the product measure of the laws of the Y_i's, $i = 1, \ldots, n$. Indeed, given $x = (x_1, \ldots, x_n) \in X$, let $t = t(x)$ achieve the supremum of $F(x)$ (start with $\mathcal{T}$ finite if necessary). Then, for every $y \in X$,

$$\begin{aligned} F(x) = \sum_{i=1}^{n} t_i x_i &\leq \sum_{i=1}^{n} t_i y_i + \sum_{i=1}^{n} |t_i||x_i - y_i| \\ &\leq F(y) + \sigma \sum_{i=1}^{n} \frac{|t_i||v_i - u_i|}{\sigma}\, \mathbf{1}_{\{x_i \neq y_i\}}. \end{aligned}$$

Hence $\sigma^{-1}F$ satisfies (4.8) with $a = a(x) = \sigma^{-1}(|t_1|, \ldots, |t_n|)$. Thus Corollary 4.7 yields the following consequence that has to be compared with (1.23) for $\mathcal{T}$ reduced to one point.

Corollary 4.8. *Let Z be as in (4.9) and denote by m_Z a median of Z. Then, for every $r \geq 0$,*

$$\mathbb{P}(\{|Z - m_Z| \geq r\}) \leq 4\,\mathrm{e}^{-r^2/4\sigma^2}.$$

Furthermore,

$$\left|\mathbb{E}(Z) - m_Z\right| \leq 4\sqrt{\pi}\,\sigma \quad \textit{and} \quad \mathrm{Var}(Z) \leq 16\sigma^2.$$

What is actually hidden behind this example are the convexity properties of the functional F. It is easy to check that if $\Omega = [0, 1]$ and if d_B is the Euclidean distance to the convex hull $B = \mathrm{Conv}(A)$ of A, then $d_B \le \mathcal{D}_A^c$. Indeed, given $x \in [0,1]^n$, for any $y^k \in A$, $\theta^k \ge 0$, $\sum_k \theta^k = 1$,

$$\begin{aligned} d_B(x) &\le \left| x - \sum_k \theta^k y^k \right| \\ &\le \bigg(\sum_{i=1}^n \Big| \sum_k \theta^k (x_i - y_i^k) \Big|^2 \bigg)^{1/2} \\ &\le \bigg(\sum_{i=1}^n \Big| \sum_k \theta^k \mathbf{1}_{\{x_i \neq y_i^k\}} \Big|^2 \bigg)^{1/2} \end{aligned}$$

since $x_i, y_i^k \in [0,1]$. The conclusion follows from the combinatorial description (4.5) of $\mathcal{D}_A^c(x)$.

Corollary 4.9. *For any product probability measure P on $[0,1]^n$, and any measurable set $A \subset [0,1]^n$,*

$$\int \mathrm{e}^{d_B^2/4} dP \le \frac{1}{P(A)}$$

where B is the convex hull of A.

Let F be a 1-Lipschitz convex function on $[0,1]^n$. Set $A = \{F \le m\}$ where $m \in \mathbb{R}$. Since F is convex, A is convex. For every $x \in [0,1]^n$ such that $d(x,A) < r$, $F(x) < m + r$. Hence, by Corollary 4.9,

$$P\big(\{F \le m\}\big) P\big(\{F \ge m + r\}\big) \le \mathrm{e}^{-r^2/4}.$$

Choosing successively as before $m = m_F$ and $m = m_F - r$ where m_F is a median of F for P, we get the following important consequence of Theorem 4.6. Alternatively, if F is convex and 1-Lipschitz, for every $x, y \in [0,1]^n$,

$$F(x) \le F(y) + \sum_{i=1}^n (x_i - y_i) \partial_i F(x) \le F(y) + d_a(x,y)$$

with $a = a(x) = (|\partial_1 F(x)|, \ldots, |\partial_n F(x)|)$ and thus F satisfies (4.8).

Corollary 4.10. *For every product probability P on $[0,1]^n$, every convex 1-Lipschitz function F on $\mathbb{R}^n$, and every $r \ge 0$,*

$$P\big(\{|F - m_F| \ge r\}\big) \le 4\,\mathrm{e}^{-r^2/4}$$

where m_F is a median of F for P.

Corollary 4.10 extends to probability measures μ_i supported on $[u_i, v_i]$, $i = 1, \ldots, n$, for functions F such that

$$\sum_{i=1}^n (v_i - u_i)^2 (\partial_i F)^2 \le 1.$$

In particular, if P is a product measure on $[u,v]^n$ and if F is convex and 1-Lipschitz on $\mathbb{R}^n$, for every $r \ge 0$,

$$P(\{|F - m_F| \ge r\}) \le 4\,\mathrm{e}^{-r^2/4(v-u)^2}. \tag{4.10}$$

By Proposition 1.8, we may replace the median by the mean up to numerical constants. The numerical constant 4 in the exponent may be improved to get close to the best possible value 2. This concentration inequality is very similar to Gaussian concentration, however with F convex. This convexity assumption cannot be omitted in general as shown by the following example. Let P be a uniform product measure on $\{0,1\}^n \subset [0,1]^n$ and assume that $n = 2k$ is even. Let the Hamming ball

$$A = \left\{x \in \{0,1\}^n ; \textstyle\sum_{i=1}^n x_i \le k\right\}$$

and $F(x) = d(x,A)$, $x \in \mathbb{R}^n$, where d is Euclidean metric. Clearly $\|F\|_{\mathrm{Lip}} = 1$ and 0 is a median of F. Now, if $\sum_{i=1}^n (2x_i - 1) \ge \rho\sqrt{k}$ for some $\rho > 0$ to be specified later, for all $y \in A$,

$$\frac{\rho}{2}\sqrt{k} \le \sum_{i=1}^n (x_i - y_i) \le \sum_{i=1}^n |x_i - y_i|^2.$$

Hence $d(x,A) \ge \sqrt{\frac{\rho}{2}}\, k^{1/4}$. It follows from the central limit theorem that

$$P\left(\left\{x ; d(x,A) \ge \sqrt{\frac{\rho}{2}}\left(\frac{n}{2}\right)^{1/4}\right\}\right) \ge P\left(\left\{x ; \frac{1}{\sqrt{n}}\sum_{i=1}^n (2x_i - 1) \ge \frac{\rho}{\sqrt{2}}\right\}\right) \ge \frac{1}{4}$$

for $\rho > 0$ sufficiently small and independent of n. Therefore concentration is not available.

Theorem 4.6 may be generalized along the same lines of proof to non-quadratic distances. One possibility is to measure the distance to $V_A(x)$ in another way. To this task, given $\beta \ge 0$, set

$$\mathcal{D}_{\beta,A}(x) = \inf_{y \in V_A(x)} \sum_{i=1}^n \tau_\beta(y_i)$$

where

$$\tau_\beta(1-u) = \beta u \log u - (1+\beta u)\log\left(\frac{1+\beta u}{1+\beta}\right), \quad u \in [0,1]. \tag{4.11}$$

The reader should observe that $\mathcal{D}_{\beta,A}$ corresponds in the preceding notation to $(\mathcal{D}_A^c)^2$ rather than to $\mathcal{D}_A^c$.

Theorem 4.11. *For every non-empty subset A of $X = \Omega^1 \times \cdots \times \Omega^n$, and every product probability measure P on Ω^n,*

$$\int \mathrm{e}^{\mathcal{D}_{\beta,A}} dP \le \frac{1}{P(A)^\beta}.$$

In particular, for every $r \ge 0$,

$$P(\{\mathcal{D}_{\beta,A} \ge r\}) \le \frac{1}{P(A)^\beta}\,\mathrm{e}^{-r}.$$

We refer to [Tal7] for a proof of Theorem 4.11 along the lines of the proof of Theorem 4.6. One striking aspect of this result is that the form of the function τ_β is exactly determined by the optimization (4.7) when dealing with the extra parameter β. We present a proof of Theorem 4.11 with the tool of information inequalities in Chapter 6.

4.3 Control by several points

The preceding convex hull approximation suggests the possibility of various notions of enlargements in product spaces that go outside the scope of distances. The following is a control by a finite number of points that is not metric.

As in the previous section, let $(\Omega_i, \sigma_i, \mu_i)$, $i = 1, \ldots, n$, be arbitrary probability spaces and let $P = \mu_1 \otimes \cdots \otimes \mu_n$ be the product measure on $X = \Omega_1 \times \cdots \times \Omega_n$. If q is an integer ≥ 2 and if $A^1, \ldots, A^q$ are subsets of X, then, for every $x = (x_1, \ldots, x_n)$ in Ω^n, let $\mathcal{D}^q(x) = \mathcal{D}^q_{A^1,\ldots,A^q}(x)$ be defined by

$$\begin{aligned}\mathcal{D}^q(x) = \inf\big\{k \geq 0; \exists\, y^1 \in A^1, \ldots, \exists\, y^q \in A^q \text{ such that} \\ \operatorname{Card}\big\{1 \leq i \leq n;\, x_i \notin \{y_i^1, \ldots, y_i^q\}\big\} \leq k\big\}.\end{aligned}$$

(We agree that $\mathcal{D}^q = \infty$ if one of the A_i's is empty.) If, for every $i = 1, \ldots, n$, $A^i = A$ for some $A \subset \Omega^n$, $\mathcal{D}^q(x) \leq k$ means that the coordinates of x may be copied, with the exception of k of them, by the coordinates of q elements in A. Using again a proof by induction on the number of coordinates, we establish the following result.

Theorem 4.12. *Under the previous notations,*

$$\int q^{\mathcal{D}^q(x)}\, dP(x) \leq \prod_{i=1}^q \frac{1}{P(A^i)}\,.$$

In particular, for every integer k,

$$P\big(\{\mathcal{D}^q \geq k\}\big) \leq q^{-k} \prod_{i=1}^q \frac{1}{P(A^i)}\,.$$

Proof. As in the proof of Theorem 4.6, we may assume that $X = \Omega^n$ is the n-fold product of a probability space (Ω, Σ, μ) with the product measure $P = \mu^n$. One first observes that if g is a function on Ω such that $\frac{1}{q} \leq g \leq 1$, then

$$\int \frac{1}{g}\, d\mu \Big(\int g d\mu\Big)^q \leq 1. \tag{4.12}$$

Indeed, since $\log u \leq u - 1$, $u \geq 0$, it suffices to show that

$$\int \frac{1}{g}\, d\mu + q \int g d\mu = \int \Big(\frac{1}{g} + qg\Big) d\mu \leq q + 1.$$

But this is obvious since $\frac{1}{u} + qu \leq q + 1$ for $\frac{1}{q} \leq u \leq 1$.

Let g_i, $i = 1, \dots, q$, be functions on Ω such that $0 \le g_i \le 1$. Applying (4.12) to g given by $\frac{1}{g} = \min(q, \min_{1\le i\le q} \frac{1}{g_i})$ yields

$$\int \min\Big(q, \min_{1\le i\le q} \frac{1}{g_i}\Big) d\mu \left(\prod_{i=1}^{q} \int g_i d\mu\right) \le 1 \tag{4.13}$$

since $g_i \le g$ for every $i = 1, \dots, q$.

We prove the theorem by induction on n. If $n = 1$, the result follows from (4.12) by taking $g_i = \mathbf{1}_{A^i}$. Assume Theorem 4.12 has been proved for n and let us prove it for $n+1$. Consider sets $A^1, \dots, A^q$ of Ω^{n+1}. For $\omega \in \Omega$, consider the sections $A^i(\omega)$, $i = 1, \dots, q$, as well as the projections B^i of A^i on Ω^n, $i = 1, \dots, q$. Note that if we set $g_i = P(A^i(\omega))/P(B^i)$ in (4.13) we get by Fubini's theorem that

$$\int \min\left(q \prod_{i=1}^{q} \frac{1}{P(B^i)}, \min_{1\le j\le q} \prod_{i=1}^{q} \frac{1}{P(C^{ij})}\right) d\mu \le \prod_{i=1}^{q} \frac{1}{P\otimes\mu(A^i)} \tag{4.14}$$

where $C^{ij} = B^i$ if $i \ne j$ and $C^{ii} = A^i(\omega)$. The basic observation is now the following: for $(x,\omega) \in \Omega^n \times \Omega$,

$$\mathcal{D}^q_{A^1,\dots,A^q}(x,\omega) \le 1 + \mathcal{D}^q_{B^1,\dots,B^q}(x)$$

and, for every $1 \le j \le q$,

$$\mathcal{D}^q_{A^1,\dots,A^q}(x,\omega) \le \mathcal{D}^q_{C^{1j},\dots,C^{qj}}(x).$$

It follows that

$$\begin{aligned}
&\int_{\Omega^{n+1}} q^{\mathcal{D}^q_{A^1,\dots,A^q}(x,\omega)} dP(x) d\mu(\omega) \\
&\quad \le \int_{\Omega^{n+1}} \min\big(q \cdot q^{\mathcal{D}^q_{B^1,\dots,B^q}(x)}, \min_{1\le j\le q} q^{\mathcal{D}^q_{C^{1j},\dots,C^{qj}}(x)}\big) dP(x) d\mu(\omega) \\
&\quad \le \int_{\Omega} \min\Big(q \int_{\Omega^n} q^{\mathcal{D}^q_{B^1,\dots,B^q}(x)} dP(x), \\
&\qquad\qquad \min_{1\le j\le q} \int_{\Omega^n} q^{\mathcal{D}^q_{C^{1j},\dots,C^{qj}}(x)} dP(x)\Big) d\mu(\omega) \\
&\quad \le \int_{\Omega} \min\left(q \prod_{i=1}^{q} \frac{1}{P(B^i)}, \min_{1\le j\le q} \prod_{i=1}^{q} \frac{1}{P(C^{ij})}\right) d\mu(\omega)
\end{aligned}$$

by the recurrence hypothesis. The conclusion follows from (4.14) and the proof of Theorem 4.12 is thus complete. □

In the applications, q is usually fixed, for example equal to 2. Theorem 4.12 then shows how to control, with a fixed subset A, arbitrary samples with an exponential decay of the probability in the number of coordinates which are neglected. Let us consider a class of functions to which this property may be applied. For a subset $I \subset \{1, \dots, n\}$, set $\Omega_I = \prod_{i\in I} \Omega_i$ and denote by Ω^* the collection of all Ω_I's, $I \subset \{1, \dots, n\}$. If $x = (x_1, \dots, x_n) \in \Omega_1 \times \dots \times \Omega_n$ and $I \subset \{1, \dots, n\}$, we denote similarly $x_I = (x_i)_{i\in I} \in \Omega_I$. A function $F : \Omega^* \to \mathbb{R}_+$ is said to be *monotone* if

$F(x_I) \leq F(x_J)$ whenever $I \subset J$. F is said to be *subadditive* if for I, J disjoint in $\{1, \ldots, n\}$ and $x_{I \cup J} \in \Omega_{I \cup J}$,

$$F(x_{I \cup J}) \leq F(x_I) + F(x_J).$$

Typical examples of monotone and subadditive functions are the following. If $\Omega_i = [0, \infty)$ for every $i = 1, \ldots, n$, set $F(x_I) = \sum_{i \in I} x_i$. If Ω_i are subsets of a normed space $(E, \| \cdot \|)$, put

$$F(x_I) = \mathbb{E}\left(\left\| \sum_{i \in I} \varepsilon_i x_i \right\| \right)$$

where ε_i are independent symmetric ± 1 Bernoulli random variables.

Corollary 4.13. *Let P be any product measure on $X = \Omega_1 \times \cdots \times \Omega_n$ and let F be monotone and subadditive on Ω^*. We denote in the same way the restriction of F to X. Let m be chosen so that (for example) $P(\{F \leq m\}) \geq \frac{1}{2}$. Then, for every integers $k, q \geq 1$, and every $r \geq 0$,*

$$P(\{F \geq qm + r\}) \leq 2^q q^{-(k+1)} + P\big(\{x; \max_{I; \mathrm{Card}(I) \leq k} F(x_I) \geq r\}\big).$$

Proof. Let $A = \{F \leq m\}$ and let $\mathcal{D}^q = \mathcal{D}^q_{A,\ldots,A}$. If $x \in \{\mathcal{D}^q \leq k\}$, there exist $y^1, \ldots, y^q$ in A such that $\mathrm{Card}\,(I) \leq k$ where

$$I = \big\{1 \leq i \leq n;\ x_i \notin \{y_i^1, \ldots, y_i^q\}\big\}.$$

Take then a partition $(J_j)_{1 \leq j \leq q}$ of $I^c = \{1, \ldots, n\} \setminus I$ such that $x_i = y_i^j$ if $i \in J_j$. Then, by subadditivity and monotonicity,

$$F(x_{I^c}) \leq \sum_{j=1}^{q} F(y^j_{J_j}) \leq \sum_{j=1}^{q} F(y^j_{\{1,\ldots,n\}}) \leq qm.$$

The conclusion immediately follows from Theorem 4.12. □

Applied to a sum of non-negative independent random variables this yields the following.

Corollary 4.14. *Let $Y_1, \ldots, Y_n$ be non-negative independent random variables on some probability space $(\Omega, \mathcal{A}, \mathbb{P})$ and set $S = Y_1 + \cdots + Y_n$. Let m be such that $\mathbb{P}(\{S \leq m\}) \geq \frac{1}{2}$ and denote by $\{Y_1^*, \ldots, Y_n^*\}$ the non-increasing rearrangement of the sample $\{Y_1, \ldots, Y_n\}$. Then, for any integers $k, q \geq 1$, and every $r \geq 0$,*

$$\mathbb{P}(\{S \geq qm + r\}) \leq 2^q q^{-(k+1)} + \mathbb{P}\left(\left\{ \sum_{i=1}^{k} Y_i^* \geq r \right\} \right).$$

If the random variables are bounded by some $C > 0$, k is usually chosen of the order of Cr so that the second term on the right-hand side of the inequality of Corollary 4.14 cancels.

Let $\mathcal{T}$ be a family of n-tuples $t = (t_1, \ldots, t_n)$, $t_i \geq 0$. It is clear that the preceding argument leading to Corollary 4.14 applies in the same way to

$$S = \sup_{t \in \mathcal{T}} \sum_{i=1}^{n} t_i Y_i$$

to yield

$$\mathbb{P}\big(\{S \geq qm + r\}\big) \leq 2^q q^{-(k+1)} + \mathbb{P}\bigg(\bigg\{\sigma \sum_{i=1}^{k} Y_i^* \geq r\bigg\}\bigg) \tag{4.15}$$

where $\sigma = \sup\{t_i; 1 \leq i \leq n, t \in \mathcal{T}\}$.

The case of random variables with arbitrary signs will be studied in this way in Chapter 7 with the tool of (probabilistic) symmetrization.

4.4 Convex infimum-convolution

In this section, we develop the tool of infimum-convolution inequalities, introduced in Section 1.6, for product measures to recover some of the results of Section 4.2. Recall from Section 1.6 that given a measurable non-negative function $\tilde{c}$ on a topological vector space X and a real-valued measurable function f on X, we denote by $Q_{\tilde{c}}f$ the infimum-convolution of f with respect to the cost $\tilde{c}(x, y) = \tilde{c}(x - y)$, $x, y \in X$,

$$Q_{\tilde{c}}f(x) = \inf_{y \in X} \big[f(y) + \tilde{c}(x - y)\big], \quad x \in X.$$

If μ is a probability measure on the Borel sets of X, and c is a non-negative measurable cost function on X, we say that μ satisfies an infimum-convolution inequality with respect to the cost $\tilde{c}$ if for all bounded measurable functions f on X,

$$\int e^{Q_{\tilde{c}}f} d\mu \int e^{-f} d\mu \leq 1. \tag{4.16}$$

We have seen in Proposition 1.18 and (1.31) how (4.16) is simply related to concentration by the fact that for every Borel set A and every $r \geq 0$,

$$1 - \mu\big(\{A + \{\tilde{c} < r\}\}\big) \leq \frac{1}{\mu(A)}\, e^{-r}. \tag{4.17}$$

As a consequence of Proposition 1.19, each time a measure μ satisfies the infimum-convolution inequality (4.16), the product measure μ^n on X^n satisfies the concentration inequality (4.16) with the cost function $\sum_{i=1}^{n} \tilde{c}(x_i)$, $x_1, \ldots, x_n \in X$.

We apply this observation in the context of the infimum-convolution inequality for convex functions. Let us say that a probability measure μ on the Borel sets of X satisfies a *convex infimum-convolution inequality* with respect to a convex cost $\tilde{c}$ if for all bounded measurable convex functions f on X,

$$\int e^{Q_{\tilde{c}}f} d\mu \int e^{-f} d\mu \leq 1. \tag{4.18}$$

It is easy to see that the convex infimum-convolution inequality also satisfies the product property of Proposition 1.19. In the following statement, $X_1, \ldots, X_n$ is a family of normed vector spaces and a point x in the product space $X = X_1 \times \cdots \times X_n$ will be denoted $x = (x_1, \ldots, x_n)$. The norms on the various X_i will be denoted in the same way by $\|\cdot\|$.

Theorem 4.15. *Let $X_1, \ldots, X_n$ be normed vector spaces and, for each $i = 1, \ldots, n$, let μ_i be a probability measure supported by a set of diameter less than or equal to 1 on X_i. Then the product probability measure $P = \mu_1 \otimes \cdots \otimes \mu_n$ satisfies the convex infimum-convolution inequality on $X = X_1 \times \cdots \times X_n$ with respect to the cost function $\tilde{c}(x) = \frac{1}{4}\sum_{i=1}^n \|x_i\|^2$, $x \in X$.*

Proof. By the product property of the convex infimum-convolution inequality, it is enough to deal with a single probability measure μ supported by a set A of diameter less than or equal to 1 on a normed space $(X, \|\cdot\|)$. Let f be a convex function on A for which we may assume without loss of generality that $\inf f(A) = 0$. Let $x \in A$, $\varepsilon > 0$ and $a \in A$ be such that $f(a) \leq \varepsilon$. For $\theta \in [0,1]$, and $y = \theta x + (1-\theta)a$, by convexity,

$$Q_{\tilde{c}} f(x) \leq f(y) + \frac{1}{4}\|x-y\|^2 \leq \theta f(x) + (1-\theta)\varepsilon + \frac{1}{4}(1-\theta)^2$$

since $\mathrm{Diam}(A) \leq 1$. Choosing optimal θ, we deduce that $Q_{\tilde{c}} f(x) \leq \psi(f(x))$ where

$$\psi(u) = \begin{cases} u - u^2 & \text{if } 0 \leq u \leq \dfrac{1}{2}, \\ \dfrac{1}{4} & \text{if } u \geq \dfrac{1}{2}. \end{cases}$$

One can check that $\mathrm{e}^{\psi(u)} \leq 2 - \mathrm{e}^{-u}$ for every $u \geq 0$. Indeed, for $0 \leq u \leq \frac{1}{2}$,

$$\frac{1}{2}\left(\mathrm{e}^{u-u^2} + \mathrm{e}^{-u}\right) = \mathrm{e}^{-u^2/2}\cosh\left(u - \frac{u^2}{2}\right) \leq \mathrm{e}^{-u^2/2}\cosh(u) \leq 1,$$

while for $u \geq \frac{1}{2}$, $\mathrm{e}^{-u} \leq 2 - \mathrm{e}^{1/4}$. It follows that

$$\int \mathrm{e}^{Q_{\tilde{c}} f} d\mu \leq 2 - \int \mathrm{e}^{-f} d\mu \leq \left(\int \mathrm{e}^{-f} d\mu\right)^{-1},$$

which proves the theorem. □

The simple minded Theorem 4.15 has strong consequences. We may indeed recover with it Corollary 4.9 and thus the useful Corollary 4.10. Indeed, let $X_i = \mathbb{R}$ for every $i = 1, \ldots, n$ and let μ_i be probability measures supported by $[0,1]$. Given a convex set A, apply Theorem 4.15 to the (convex) function f equal to 0 on A and to $+\infty$ outside. Since $Q_{\tilde{c}} f = \frac{1}{4} d(\cdot, A)^2$, the claim follows.

4.5 The exponential distribution

In this section, we consider a special concentration property of the exponential distribution. Let ν^n be the product measure on $\mathbb{R}^n$ when each factor is endowed with the measure ν of density $\frac{1}{2}\mathrm{e}^{-|x|}$ with respect to Lebesgue measure. Denote by $\mathcal{B}^n = \mathcal{B}_2^n$ the Euclidean (open) unit ball and by $\mathcal{B}_1^n$ the ℓ^1-unit ball in $\mathbb{R}^n$, i.e.,

$$\mathcal{B}_1^n = \Big\{ x = (x_1, \ldots, x_n) \in \mathbb{R}^n; \sum_{i=1}^n |x_i| < 1 \Big\}.$$

Theorem 4.16. *For every Borel set A in $\mathbb{R}^n$ and every $r > 0$,*

$$1 - \nu^n\big(A + 6\sqrt{r}\,\mathcal{B}_2^n + 9r\mathcal{B}_1^n\big) \leq \frac{1}{\nu^n(A)}\,\mathrm{e}^{-r}.$$

A striking feature of Theorem 4.16 is that it may be used to improve some aspects of concentration for Gaussian measures especially for cubes. Consider indeed the increasing map $\Psi : \mathbb{R} \to \mathbb{R}$ that transforms ν into the one-dimensional canonical Gaussian measure γ. It is a simple matter to check that

$$\big|\Psi(x) - \Psi(y)\big| \leq K \min\big(|x-y|, |x-y|^{1/2}\big), \quad x, y \in \mathbb{R}, \tag{4.19}$$

for some numerical constant $K > 0$. The map $\varphi = \Psi^{\otimes n} : \mathbb{R}^n \to \mathbb{R}^n$ defined by $\varphi(x) = (\Psi(x_i))_{1 \leq i \leq n}$ transforms ν^n into γ^n. Consider now a Borel set A of $\mathbb{R}^n$ such that $\gamma^n(A) \geq \frac{1}{2}$. Then, by Theorem 4.16,

$$\begin{aligned}\gamma^n\big(\varphi(\varphi^{-1}(A) + 6\sqrt{r}\,\mathcal{B}_2^n + 9r\mathcal{B}_1^n)\big) & \\ &= \nu^n\big(\varphi^{-1}(A) + 6\sqrt{r}\,\mathcal{B}_2^n + 9r\mathcal{B}_1^n\big) \\ &\geq 1 - 2\,\mathrm{e}^{-r}.\end{aligned}$$

However, it follows from (4.19) that

$$\varphi\big(\varphi^{-1}(A) + 6\sqrt{r}\,\mathcal{B}_2^n + 9r\mathcal{B}_1^n\big) \subset A + K'\sqrt{r}\,\mathcal{B}_2^n.$$

Thus Theorem 4.16 implies (within numerical constants) concentration for γ^n. To actually illustrate the improvement for some classes of sets, let

$$A = \big\{x \in \mathbb{R}^n; \max_{1 \leq i \leq n} |x_i| \leq m\big\}$$

where $m = m(n)$ is chosen so that $\gamma^n(A) \geq \frac{1}{2}$ (and hence $m(n)$ is of order $\sqrt{\log n}$). Then, when $r \geq 1$ is very small compared to $\log n$, it is easily seen that actually

$$\begin{aligned}\varphi\big(\varphi^{-1}(A) + \sqrt{r}\,\mathcal{B}_2^n + r\mathcal{B}_1^n\big) &\subset A + K_1\bigg(\frac{6\sqrt{r}}{\sqrt{\log n}}\,\mathcal{B}_2^n + \frac{9r}{\sqrt{\log n}}\,\mathcal{B}_1^n\bigg) \\ &\subset A + K_2\sqrt{\frac{r}{\log n}}\,\sqrt{r}\,\mathcal{B}_2^n.\end{aligned}$$

To establish Theorem 4.16, we show that the exponential distribution satisfies the infimum-convolution inequality (4.16) for a cost that gives rise to the enlargement $6\sqrt{r}\,\mathcal{B}_2^n + 9r\mathcal{B}_1^n$ through (4.17). By Proposition 1.19, it will be enough to deal with the dimension one. By a further symmetrization argument, we need only consider the one-sided exponential measure with density e^{-x} with respect to Lebesgue measure on $\mathbb{R}_+$.

Proposition 4.17. *The probability measure with density* e^{-x} *on* $\mathbb{R}_+$ *satisfies the infimum-convolution inequality with cost function* $\tilde{c}(x-y)$ *on* $\mathbb{R}\times\mathbb{R}$ *where*

$$\tilde{c}(x) = \begin{cases} \dfrac{1}{18}\,x^2 & \text{if } |x| \le 2\,, \\ \dfrac{2}{9}\,\big(|x|-1\big) & \text{if } |x| > 2\,. \end{cases}$$

Proof. Let f be a bounded measurable function on $\mathbb{R}_+$. Let

$$J_0 = \int_0^\infty \mathrm{e}^{-f(x)-x}dx \quad \text{and} \quad J_1 = \int_0^\infty \mathrm{e}^{Q_{\tilde{c}}f(y)-y}dy.$$

For $0<t<1$, define $x(t)$ and $y(t)$ by the relations

$$\int_0^{x(t)} \mathrm{e}^{-f(x)-x}dx = tJ_0 \quad \text{and} \quad \int_0^{y(t)} \mathrm{e}^{Q_{\tilde{c}}f(y)-y}dy = tJ_1.$$

By differentiation,

$$x'(t) = J_0\,\mathrm{e}^{f(x(t))+x(t)} \quad \text{and} \quad y'(t) = J_1\,\mathrm{e}^{-Q_{\tilde{c}}f(y(t))+y(t)}.$$

Here y' is the usual derivative while x' is understood in the weak sense of the Sobolev space H_1. Since

$$Q_{\tilde{c}}f\big(y(t)\big) \le f\big(x(t)\big) + c\big(x(t)-y(t)\big),$$

we get

$$y'(t) \ge J_1\,\mathrm{e}^{-f(x(t))-\tilde{c}(x(t)-y(t))+y(t)}.$$

Now let $z(t) = \frac{1}{2}\,(x(t)+y(t)) - \tilde{c}(x(t)-y(t))$. We have

$$z'(t) = \Big(\tfrac{1}{2} - \tilde{c}'\big(x(t)-y(t)\big)\Big)x'(t) + \Big(\tfrac{1}{2} + \tilde{c}'\big(x(t)-y(t)\big)\Big)y'(t).$$

Now $|\tilde{c}'| \le \frac{1}{2}$. Writing x for $x(t)$ and y for $y(t)$, and using the inequality $\lambda u + \lambda^{-1}v \ge 2\sqrt{uv}$ for $\lambda = \mathrm{e}^{f(x)}$, we get

$$\begin{aligned} z'(t) &\ge \big(\tfrac{1}{2} - \tilde{c}'(x-y)\big)J_0\,\mathrm{e}^{x+f(x)} \\ &\quad + \big(\tfrac{1}{2} + \tilde{c}'(x-y)\big)J_1\,\mathrm{e}^{-\tilde{c}(x-y)+y-f(x)} \\ &\ge \sqrt{1-4\tilde{c}'(x-y)^2}\,\sqrt{J_0J_1}\,\mathrm{e}^{\frac{1}{2}(x+y)-\frac{1}{2}\tilde{c}(x-y)} \\ &\ge \sqrt{J_0J_1}\,\mathrm{e}^{z(t)}\sqrt{1-4\tilde{c}'(x-y)^2}\,\mathrm{e}^{\frac{1}{2}\tilde{c}(x-y)}. \end{aligned}$$

Now, $(1-4\tilde{c}'(u)^2)\,\mathrm{e}^{\tilde{c}(u)} \ge 1$ for every $u\in\mathbb{R}$. Since $\tilde{c}$ is even, it is enough to check the latter for $u \ge 0$. For $u \ge 2$, $\tilde{c}'$ is constant and $\tilde{c}$ is increasing. For $0 \le u \le 2$, it reduces to the elementary inequality $\mathrm{e}^{-v/18} \leq 1 - (4v/81)$ for $0 \leq v \leq 4$. As a consequence, it follows that

$$\mathrm{e}^{-z(t)} z'(t) \geq \sqrt{J_0 J_1},$$

which, after integration between 0 and 1, yields the result. Proposition 4.17 is established. □

To complete the proof of Theorem 4.16, observe first that Proposition 4.17 holds, up to numerical constants, for the two-sided exponential distribution with the cost function

$$\tilde{c}(x) = \begin{cases} \dfrac{1}{36}\, x^2 & \text{if } |x| \leq 4\,, \\ \dfrac{2}{9}\,(|x| - 2) & \text{if } |x| > 4\,. \end{cases} \tag{4.20}$$

By Proposition 1.19, the product ν^n of the two-sided exponential distribution on $\mathbb{R}^n$ satisfies the infimum-convolution inequality with the cost $\sum_{i=1}^n \tilde{c}(x_i)$, $x = (x_1, \ldots, x_n) \in \mathbb{R}^n$. It is then a mere exercise to check that

$$\Big\{ x \in \mathbb{R}^n ; \sum_{i=1}^n \tilde{c}(x_i) < r \Big\} \subset 6\sqrt{r}\, \mathcal{B}_2^n + 9r \mathcal{B}_1^n$$

from which the conclusion follows by (4.17).

As in Chapter 1, the concentration inequality on sets of Theorem 4.16 may be translated equivalently to functions. This is the content of the next proposition.

Proposition 4.18. *Let F be a real-valued function on $\mathbb{R}^n$ such that $\|F\|_{\mathrm{Lip}} \leq a$ and such that its Lipschitz coefficient with respect to the ℓ^1-metric is less than or equal to b, that is,*

$$\big|F(x) - F(y)\big| \leq b \sum_{i=1}^n |x_i - y_i|, \quad x, y \in \mathbb{R}^n.$$

Then, for every $r \geq 0$,

$$\nu^n\big(\{F \geq M + r\}\big) \leq K \exp\bigg(-\frac{1}{K} \min\Big(\frac{r}{b}, \frac{r^2}{a^2}\Big)\bigg) \tag{4.21}$$

for some numerical constant $K > 0$ where M is either a median of F for ν^n or its mean. Conversely, if the latter holds, the concentration result of Theorem 4.16 holds.

By Rademacher's theorem, the hypotheses on F are equivalent to saying that F is almost everywhere differentiable with

$$\sum_{i=1}^n |\partial_i F|^2 \leq a^2 \quad \text{and} \quad \max_{1 \leq i \leq n} |\partial_i F| \leq b$$

almost everywhere. The inequality of Proposition 4.18 extends in the appropriate sense the classical case of linear functions F (sums of exponential random variables) with a quadratic exponential growth for the small values of r and a linear one for the large values, and may be shown to be sharp in this case.

Proof. Apply first Theorem 4.16 to $A = \{F \leq m_F\}$ where m_F is a median of F for ν^n and note that by the Lipschitz bounds on F,

$$A + 6\sqrt{r}\,\mathcal{B}_2^n + 9r\mathcal{B}_1^n \subset \{F \leq m_F + 6a\sqrt{r} + 9br\}.$$

The positive solution s to the equation $6a\sqrt{r}+9br = s$ behaves like $s \approx \min\big(\frac{s}{b}, \frac{s^2}{a^2}\big)$. This yields the inequality of the proposition for the median. Using a routine argument (cf. Proposition 1.8), the deviation inequality from either the median or the mean are equivalent up to numerical constants. Conversely, for $A \subset \mathbb{R}^n$ and $x = (x_1, \dots, x_n) \in \mathbb{R}^n$, set

$$F_A(x) = \inf_{z\in A} \sum_{i=1}^n \min\big(|x_i - z_i|, |x_i - z_i|^2\big).$$

For $r > 0$, set further $F = \min(F_A, r)$. Then $\sum_{i=1}^n |\partial_i F|^2 \leq 4r$ and $\max_{1\leq i\leq n} |\partial_i F|$ ≤ 2 almost everywhere. Indeed, it is enough to prove this result for $G = \min(G_z, r)$ for every fixed z where

$$G_z(x) = \sum_{i=1}^n \min\big(|x_i - z_i|, |x_i - z_i|^2\big).$$

Now, almost everywhere, and for every $i = 1, \dots, n$, $|\partial_i G_z(x)| \leq 2|x_i - z_i|$ if $|x_i - z_i| \leq 1$ whereas $|\partial_i G_z(x)| \leq 1$ if $|x_i - z_i| > 1$. Therefore, $\max_{1\leq i\leq n} |\partial_i G_z(x)| \leq 2$ and

$$\sum_{i=1}^n \big|\partial_i G_z(x)\big|^2 \leq 4\sum_{i=1}^n \min\big(|x_i - z_i|, |x_i - z_i|^2\big) = 4G_z(x)$$

which yields the announced claim. If $\nu^n(A) \geq \frac{1}{2}$, 0 is a median of F so that, by the hypothesis with M the median,

$$\nu^n\big(\{F_A \geq r\}\big) = \nu^n\big(\{F \geq r\}\big) \leq \mathrm{e}^{-r/4K}.$$

Since $\{F_A \leq r\} \subset A + \sqrt{r}\,\mathcal{B}_2^n + r\mathcal{B}_1^n$, this amounts to the inequality of Theorem 4.16. □

In the final corollary, we apply Theorem 4.16 to concentration of products of the measure ν_p with density $c_p\,\mathrm{e}^{-|x|^p}$, $1 \leq p < \infty$, with respect to Lebesgue measure on $\mathbb{R}$. We recall the ℓ^p-unit ball $\mathcal{B}_p^n$, $1 \leq p < \infty$, in $\mathbb{R}^n$,

$$\mathcal{B}_p^n = \bigg\{x = (x_1, \dots, x_n) \in \mathbb{R}^n; |x|_p = \bigg(\sum_{i=1}^n |x_i|^p\bigg)^{1/p} < 1\bigg\}$$

and

$$\mathcal{B}_\infty^n = \big\{x = (x_1, \dots, x_n) \in \mathbb{R}^n; |x|_\infty = \max_{1\leq i\leq n} |x_i| < 1\big\}.$$

Denote by ν_p^n the product measure of ν_p on $\mathbb{R}^n$.

Theorem 4.19. *There is a constant $K > 0$ only depending on $1 \le p < \infty$ such that for any Borel set A in $\mathbb{R}^n$ and any $r > 0$,*

$$1 - \nu_p^n\big(A + \sqrt{r}\,\mathcal{B}_2^n + r^{1/p}\mathcal{B}_p^n\big) \le \frac{1}{\nu_p^n(A)}\,\mathrm{e}^{-r/K}.$$

Proof. For $p = 1$, this is the content of Theorem 4.16. We show how this case implies the other ones. As for the comparison with Gaussian measures, consider the increasing transformation Ψ_p from $\mathbb{R}$ to $\mathbb{R}$ that maps ν_1 into ν_p. Denote then by $\varphi_p = \Psi_p^{\otimes n}$ the map from $\mathbb{R}^n$ into itself given by $\varphi_p(x) = (\Psi_p(x_i))_{1 \le i \le n}$, $x = (x_1, \dots, x_n) \in \mathbb{R}^n$; φ_p transforms ν_1^n into ν_p^n. If $\nu_p^n(A) \ge \frac{1}{2}$, then $\nu_1^n(\varphi_p^{-1}(A)) \ge \frac{1}{2}$. By Theorem 4.16,

$$\nu_1^n\big(\varphi_p^{-1}(A) + 6\sqrt{r}\,\mathcal{B}_2^n + 9r\mathcal{B}_1^n\big) \ge 1 - 2\,\mathrm{e}^{-r}$$

for every $r > 0$. Thus it suffices to show that

$$\varphi_p\big(\varphi_p^{-1}(A) + 6\sqrt{r}\,\mathcal{B}_2^n + 9r\mathcal{B}_1^n\big) \subset A + K\big(\sqrt{r}\,\mathcal{B}_2^n + r^{1/p}\mathcal{B}_p^n\big). \tag{4.22}$$

To this task, we need the following simple but a bit technical estimates we leave to the reader (see [Tal6]). First, for all $x, y \in \mathbb{R}$,

$$\big|\Psi_p(x) - \Psi_p(y)\big| \le K \min\big(|x-y|, |x-y|^{1/p}\big). \tag{4.23}$$

(Compare with (4.19).) Here and below, $K > 0$ denotes a constant only depending on p and possibly changing from line to line. Second, define $\vartheta_p(u) = u^2$ for $|u| \le 1$ and $\vartheta_p(u) = u^p$ for $|u| \ge 1$ and set for $r > 0$,

$$V_p(r) = \Big\{x \in \mathbb{R}^n;\ \sum_{i=1}^n \vartheta_p(x_i) < r\Big\}.$$

Then

$$\sqrt{r}\,\mathcal{B}_2^n + r\mathcal{B}_1^n \subset 12V_1(r), \tag{4.24}$$

while for every p,

$$V_p(r) \subset \sqrt{r}\,\mathcal{B}_2^n + r^{1/p}\mathcal{B}_p^n. \tag{4.25}$$

Therefore, by (4.24) and (4.25), in order to prove (4.22), it suffices to show that for every $r > 0$,

$$\varphi_p\big(\varphi_p^{-1}(A) + 2V_1(r)\big) \subset A + V_p(Kr).$$

Consider $y \in \varphi_p^{-1}(A) + 2V_1(r)$. Thus $y = \varphi_p^{-1}(x) + z$ where $x \in A$ and $\sum_{i=1}^n \vartheta_p(\frac{z_i}{2}) < r$. By (4.23),

$$\big|\Psi_p(y_i) - x_i\big| \le K \min\big(|z_i|, |z_i|^{1/p}\big)$$

for every i. If $|z_i| \le 1$,

$$\vartheta_p\Big(\frac{\Psi_p(y_i) - x_i}{K}\Big) \le z_i^2$$

while if $|z_i| \ge 1$, since ϑ_p increases,

$$\vartheta_p\Big(\frac{\Psi_p(y_i) - x_i}{K}\Big) \le |z_i|.$$

Now $\vartheta_1(u) = \min(u^2, |u|) \le K\vartheta_1(\frac{u}{2})$ so that

$$\vartheta_p\Big(\frac{\Psi_p(y_i) - x_i}{K}\Big) \le \vartheta_1(z_i) \le K\vartheta_1\big(\tfrac{z_i}{2}\big)$$

and

$$\sum_{i=1}^n \vartheta_p\Big(\frac{\Psi_p(y_i) - x_i}{K}\Big) \le Kr.$$

The proof is easily completed. □

As in Proposition 4.18, the concentration result of Theorem 4.19 may be translated equivalently to functions.

It should be noted that the arguments developed in Propositions 2.9 and 2.10 transferring concentration for Gaussian measures to concentration on the Euclidean ball and sphere may be applied similarly starting from Theorem 4.16 (or rather Proposition 4.18) to yield a concentration result on the ℓ^1-unit ball or sphere in $\mathbb{R}^n$. Indeed, if x is distributed according to ν_1^n, then $x/|x|_1$ is distributed according to the normalized surface measure on the sphere of ℓ^1 in $\mathbb{R}^n$. We may thus reach in this way concentration for uniform measure on the unit ball, or sphere, of ℓ^1, hence including the value $p = 1$ in (2.27).

Proposition 4.20. *If μ is uniformly distributed on the unit sphere, or ball, of ℓ^p in $\mathbb{R}^n$, $1 \le p \le 2$,*

$$\alpha_{(\mathcal{B}_p^n, |\cdot|_p, \mu)}(r) \le C\,\mathrm{e}^{-cnr^2}, \quad r > 0,$$

where $C, c > 0$ are constants depending only on p.

We refer to [AR-V], [Sche4] for details and proofs. In particular, on the basis of Theorem 4.19, the result may be obtained simultaneously for every $1 \le p \le 2$. Recall that in the case of $1 < p < 2$, the relevant measure μ is not the surface measure but the one induced from Lebesgue measure on the full ball defined in (2.26).

In another direction, the paper [S-Z2] deals with Lipschitz functions with respect to the Euclidean metric.

Proposition 4.21. *If μ is uniformly distributed on the unit ball, or sphere, of ℓ^p in $\mathbb{R}^n$, $1 \le p \le 2$,*

$$\alpha_{(\mathcal{B}_p^n, |\cdot|_2, \mu)}(r) \le C\,\mathrm{e}^{-cnr^p}, \quad r > 0,$$

where $C, c > 0$ are constants depending only on p.

Notes and Remarks

This chapter presents various views on concentration for product measures. Lemma 4.1 goes back to [Az]. The generality of the martingale method as presented in Section 4.1 was understood by G. Schechtman [Sche1] (from which most of the results of this section are taken; see also [M-S]) after the important step by B. Maurey [Mau1] on the symmetric group (Corollary 4.3). Its first occurrence however goes back to the Yurinskii method of Corollary 4.5 [Yu1], [Yu2] to bound norms of sums of independent random vectors from their mean in probability in Banach spaces (cf. [Le-T]). Optimal concentration functions on the discrete cube via the martingale method are discussed in [MD1] and [Tal7]. The notes [MD2] form a complete survey of the applications of the bounded difference martingale method to algorithmic discrete mathematics. Recent developments of the methods towards fluctuations results are due to O. Catoni [Ca].

It soon became apparent that the Laplace approach would not yield optimal results when applied to more general functionals. In [Tal1], M. Talagrand established Corollary 4.9 for a uniform measure on the discrete cube. W. Johnson and G. Schechtman [J-S2] extended the argument to all measures on the discrete cube that prompted M. Talagrand to the abstract formulation presented here as Section 4.2. There is an analogue of Theorem 4.6 on the symmetric group [Tal7], extended recently by C. McDiarmid [MD3], and presented in Section 8.2 below.

Motivated by questions in probability in Banach spaces (cf. Chapter 7), M. Talagrand [Tal2] established simultaneously Theorem 4.12. The first proof uses delicate rearrangement and symmetrization techniques of isoperimetric flavor. The short proof presented here appears in [Tal7].

The memoir [Tal7] by M. Talagrand is a landmark paper on concentration in product spaces, which presents definitive results with applications to a number of questions on norms of sums of independent random vectors and discrete algorithmic probabilities. See also [Tal8] for an introduction. The results of Sections 4.2 and 4.3 are taken from there and some applications are developed in Chapters 7 and 8. Penalty versions of the results presented here with applications to distributions with unbounded supports are also investigated in [Tal7].

The application of infimum-convolutions to concentration properties of product measures was emphasized by B. Maurey in [Mau2]. The results of Section 4.4 are taken from this reference.

It was again the merit of M. Talagrand [Tal3] to emphasize the concentration properties of the exponential distribution. The exposition of Section 4.5 is entirely taken from the nice approach developed by B. Maurey [Mau2]. The method is already implicitly present in Talagrand's work, which reaches moreover some isoperimetric statements. Theorem 4.19 is taken from [Tal6]. The papers [M-P] and [S-Z1] develop the crucial argument to transfer the measures ν_p^n to uniform measures on ℓ^p-balls and spheres. Applications of this principle to concentration on the ℓ^p-balls, $1 \leq p \leq 2$, (Propositions 4.20 and 4.21) appear in [S-S1], [S-Z1], [S-Z2], [AR-V], [Sche4]. These results may be used in various embedding questions (cf. [Sche5]).

5. ENTROPY AND CONCENTRATION

In this chapter, we illustrate how normal concentration properties may follow from a functional logarithmic Sobolev inequality. This is in parallel with the exponential concentration under spectral bounds investigated in Section 3.1. Although rather elementary, this observation is a powerful scheme which allows us to both establish some new concentration inequalities and recover several of the concentration results in product spaces presented in the previous chapter due to the basic product property of entropy. The first section presents the logarithmic Sobolev inequalities introduced by L. Gross [Gros1] and develops the basic scheme from logarithmic Sobolev inequalities to concentration that is going back to some unpublished observation by I. Herbst. We next investigate entropy of product measures and their applications to concentration in product spaces. Section 5.3 analyzes, with the tool of logarithmic Sobolev inequalities, the concentration properties of the exponential distribution while section 5.4 is devoted to some discrete analogues. The last section describes related covariance identities that also entail measure concentration.

5.1 Logarithmic Sobolev inequalities and concentration

We present in this section the Herbst argument leading to measure concentration from a logarithmic Sobolev inequality.

Given a probability measure μ on some measurable space (Ω, Σ), for every non-negative measurable function f on (Ω, Σ), define its *entropy* as

$$\mathrm{Ent}_\mu(f) = \int f \log f d\mu - \int f d\mu \log \int f d\mu$$

if $\int f \log(1+f) d\mu < \infty$, and $+\infty$ if not. Note that $\mathrm{Ent}_\mu(f) \geq 0$ by Jensen's inequality and that entropy is homogeneous of degree 1.

We introduce the concept of *logarithmic Sobolev inequality*. To avoid some technical questions, let us consider first the case of the Euclidean space $\mathbb{R}^n$. A probability measure μ on the Borel sets of $\mathbb{R}^n$ is said to satisfy a logarithmic Sobolev inequality if for some constant $C > 0$ and all smooth enough functions f on $\mathbb{R}^n$,

$$\mathrm{Ent}_\mu(f^2) \leq 2C \int |\nabla f|^2 d\mu. \tag{5.1}$$

Here ∇f denotes the usual gradient of f and $|\nabla f|$ its Euclidean length. Hereinafter, by smooth we understand enough regularity so that the various terms in (5.1) make sense.

The logarithmic Sobolev inequality is to be compared with the Poincaré inequality (or spectral gap inequality) of Section 3.1,

$$\mathrm{Var}_\mu(f) \le C \int |\nabla f|^2 d\mu \tag{5.2}$$

where we recall that

$$\mathrm{Var}_\mu(f) = \int f^2 d\mu - \left(\int f d\mu \right)^2$$

is the variance of a (square integrable) function f. Applying (5.1) to $1 + \varepsilon f$ and letting $\varepsilon \to 0$ shows by a standard Taylor expansion that the logarithmic Sobolev inequality (5.1) is a stronger statement than the Poincaré inequality (5.2). This observation moreover motivates the choice of the normalization $2C$ in (5.1). The basic example of the canonical Gaussian measure γ on $\mathbb{R}^n$ satisfies (5.1) (and thus also (5.2)) with $C = 1$. As we already mentioned in Chapter 3, the constant 1 in the Poincaré inequality for γ may also be described as the eigenvalue of the first Hermite polynomial in the orthogonal decomposition of $\mathrm{L}^2(\gamma)$. While Poincaré inequalities are spectral, this is not the case for logarithmic Sobolev inequalities, which do not usually follow from eigenfunction expansions.

Logarithmic Sobolev inequalities are part of the family of classical Sobolev inequalities. In terms of Sobolev embeddings, under a logarithmic Sobolev inequality, functions in H^1 do not belong necessarily to some L^p-space with $p > 2$, but to the Orlicz space $\mathrm{L}^2 \log \mathrm{L}$. This embedding is optimal for the basic example of Gaussian measures. On the other hand, no constant depending on the dimension arises in the logarithmic Sobolev inequality for Gaussian measures. This is one fundamental aspect of the infinite dimensional character of logarithmic Sobolev inequalities that will be exploited here toward dimension free concentration.

Although we already provided a number of arguments leading to concentration of Gaussian measures, we present, for the matter of completeness, a proof of the logarithmic Sobolev inequality for Gaussian measures (that will lead below to concentration). The argument relies on the semigroup tools of Section 2.3.

Theorem 5.1. *For every smooth enough function f on $\mathbb{R}^n$,*

$$\mathrm{Ent}_\gamma(f^2) \le 2 \int |\nabla f|^2 d\gamma.$$

Proof. Recall from Section 2.3 the second-order differential operator $\mathrm{L} = \Delta - x \cdot \nabla$ on $\mathbb{R}^n$ with associated semigroup $(P_t)_{t \ge 0}$ called the Ornstein-Uhlenbeck semigroup (cf. [Bak1]). In this particular example, $(P_t)_{t \ge 0}$ admits an explicit integral representation as

$$P_t f(x) = \int f\big(\mathrm{e}^{-t} x + (1 - \mathrm{e}^{-2t})^{1/2} y\big) d\gamma(y), \quad t \ge 0,\ x \in \mathbb{R}^n. \tag{5.3}$$

Let f be smooth and non-negative on $\mathbb{R}^n$. To be more precise, we take f smooth and such that $\varepsilon \le f \le 1/\varepsilon$ for some $\varepsilon > 0$, which we set to 0 at the end of the argument. Since $P_0 f = f$ and $\lim_{t \to \infty} P_t f = \int f d\gamma$, write

$$\mathrm{Ent}_\gamma(f) = - \int_0^\infty \frac{d}{dt} \left(\int P_t f \log P_t f d\gamma \right) dt.$$

By the chain rule formula and integration by parts for L (2.33),

$$\begin{aligned}\frac{d}{dt}\int P_t f \log P_t f d\gamma &= \int \mathrm{L} P_t f \log P_t f d\gamma + \int \mathrm{L} P_t f d\gamma \\ &= -\int \frac{|\nabla P_t f|^2}{P_t f} d\gamma\end{aligned}$$

since γ is invariant under the action of P_t and thus $\int \mathrm{L} P_t f d\gamma = 0$. Now, by the integral representation (5.3), for every $t \geq 0$,

$$\nabla P_t f = \mathrm{e}^{-t} P_t(\nabla f)$$

and thus

$$\left|\nabla P_t f\right| \leq \mathrm{e}^{-t} P_t\big(|\nabla f|\big). \tag{5.4}$$

By the Cauchy-Schwarz inequality for P_t,

$$P_t\big(|\nabla f|\big)^2 \leq P_t f\, P_t\Big(\frac{|\nabla f|^2}{f}\Big).$$

Summarizing,

$$\mathrm{Ent}_\gamma(f) \leq \int_0^\infty \mathrm{e}^{-2t}\left(\int P_t\Big(\frac{|\nabla f|^2}{f}\Big) d\mu\right) dt = \frac{1}{2}\int \frac{|\nabla f|^2}{f} d\mu$$

by invariance. By the change of f into f^2 the theorem is established. □

The preceding proof works similarly for all log-concave measures $d\mu = \mathrm{e}^{-U} dx$ such that $\mathrm{Hess}\, U(x) \geq c\,\mathrm{Id} > 0$ for some $c > 0$ uniformly in $x \in \mathbb{R}^n$.

Theorem 5.2. *Let $d\mu = \mathrm{e}^{-U} dx$ where, for some $c > 0$, $\mathrm{Hess}\, U(x) \geq c\,\mathrm{Id}$ uniformly in $x \in \mathbb{R}^n$. Then for all smooth functions f on $\mathbb{R}^n$,*

$$\mathrm{Ent}_\mu(f^2) \leq \frac{2}{c}\int |\nabla f|^2 d\mu.$$

In order to establish Theorem 5.2, one should however note that the argument above (cf. (5.4)) requires to improve (2.32) into

$$\left|\nabla P_t f\right| \leq \mathrm{e}^{-ct} P_t\big(|\nabla f|\big)$$

where, as in Section 2.3, $(P_t)_{t\geq 0}$ is the semigroup with generator $\mathrm{L} = \Delta - \nabla U \cdot \nabla$ (cf. [Bak2], [Le6] for a proof of Theorem 5.2 along these lines). Alternatively, following the proof of Theorem 5.1, set (with the corresponding notations)

$$\phi(t) = \int \frac{|\nabla P_t f|^2}{P_t f} d\mu, \quad t \geq 0.$$

Check then that $\phi'(t) \leq -2c\phi(t)$, $t \geq 0$, as a consequence of (2.34) applied to $\log P_t f$. Hence $\phi(t) \leq \mathrm{e}^{-2ct}\phi(0)$, $t \geq 0$, from which the conclusion follows (cf. [B-E], [Bak1], [Le7]). As we have seen in Theorem 2.7, such measures satisfy a

Gaussian type isoperimetric inequality. It may be shown that this isoperimetric inequality actually implies the logarithmic Sobolev inequality (cf. [Le5]).

The proof of Theorems 5.1 and 5.2 further extends to (compact) Riemannian manifolds (X, g) with a strictly positive lower bound $c = c(X) > 0$ on the Ricci curvature. As for the first non-trivial eigenvalue $\lambda_1 = \lambda_1(X)$ of the Laplace operator Δ on a compact Riemannian manifold (X, g), we may define formally the logarithmic Sobolev constant of Δ or of the normalized Riemannian measure μ on (X, g) as the constant $\rho_0 = \rho_0(X)$ such that

$$\rho_0 \mathrm{Ent}_\mu(f^2) \le 2 \int f(-\Delta f) d\mu = 2 \int |\nabla f|^2 d\mu \tag{5.5}$$

holds for all smooth functions f on (X, g). By the Sobolev embedding theorem, it is known [Rot1] that $\rho_0 > 0$ on any compact manifold, and furthermore as above that $\rho_0 \le \lambda_1$. The proof of Theorem 5.1 thus extends, as for Theorem 5.2, to show that

$$\lambda_1 \ge \rho_0 \ge c(X). \tag{5.6}$$

In analogy with the Lichnerowicz lower bound $\lambda_1 \ge \frac{nc(X)}{n-1}$ [Cha1], [G-H-L], this may actually be improved [B-E] together with the dimension n of X as

$$\lambda_1 \ge \rho_0 \ge \frac{nc(X)}{n-1}.$$

In particular,

$$\mathrm{Ent}_{\sigma^n}(f^2) \le \frac{2}{n} \int |\nabla f|^2 d\sigma^n \tag{5.7}$$

on the standard n-sphere $\mathbb{S}^n$, including $n = 1$. We refer to [Bak1], [Le6] for further details on these aspects and for the proofs of (5.6) and (5.7).

We now present the *Herbst argument* from a logarithmic Sobolev inequality to concentration. The principle is similar to the application of spectral properties to concentration presented in Section 3.1, but logarithmic Sobolev inequalities allow us to reach normal concentration. To begin with, let us consider a probability measure μ on the Borel sets of $\mathbb{R}^n$. Assume that μ satisfies the logarithmic Sobolev inequality (5.1). We will then show that

$$\mathrm{E}_\mu(\lambda) \le \mathrm{e}^{C\lambda^2/2}, \quad \lambda \ge 0,$$

where we recall that E_μ is the Laplace functional of μ (Section 1.6). Thus let F be a smooth bounded Lipschitz function on $\mathbb{R}^n$ such that $\int F d\mu = 0$. In particular, since F is assumed to be regular enough, we can have that $|\nabla F| \le \|F\|_{\mathrm{Lip}}$ at every point. We apply (5.1) to $f^2 = \mathrm{e}^{\lambda F - C\lambda^2 \|F\|_{\mathrm{Lip}}^2/2}$ for every $\lambda \in \mathbb{R}$. We have

$$\begin{aligned} \int |\nabla f|^2 d\mu &= \frac{\lambda^2}{4} \int |\nabla F|^2 \mathrm{e}^{\lambda F - C\lambda^2 \|F\|_{\mathrm{Lip}}^2/2} d\mu \\ &\le \frac{\lambda^2}{4} \int \mathrm{e}^{\lambda F - C\lambda^2 \|F\|_{\mathrm{Lip}}^2/2} d\mu. \end{aligned}$$

Setting $\Lambda(\lambda) = \int e^{\lambda F - C\lambda^2 \|F\|_{\mathrm{Lip}}^2/2} d\mu$, $\lambda \in \mathbb{R}$, by the definition of entropy,

$$\int \Big[\lambda F - \frac{C}{2}\,\lambda^2 \|F\|_{\mathrm{Lip}}^2\Big] e^{\lambda F - C\lambda^2 \|F\|_{\mathrm{Lip}}^2/2} d\mu - \Lambda(\lambda)\log\Lambda(\lambda) \leq \frac{C}{2}\,\lambda^2 \|F\|_{\mathrm{Lip}}^2 \Lambda(\lambda).$$

In other words,

$$\lambda\Lambda'(\lambda) \leq \Lambda(\lambda)\log\Lambda(\lambda), \quad \lambda \in \mathbb{R}.$$

If $H(\lambda) = \frac{1}{\lambda}\log\Lambda(\lambda)$ (with $H(0) = \Lambda'(0)/\Lambda(0) = \int F d\mu = 0$), $\lambda \in \mathbb{R}$, then $H'(\lambda) \leq 0$ for every λ. Therefore, H is non-increasing and thus $\Lambda(\lambda) \leq 1$ for every λ, that is,

$$\int e^{\lambda F} d\mu \leq e^{C\lambda^2 \|F\|_{\mathrm{Lip}}^2/2}. \tag{5.8}$$

Replacing F by a smooth convolution, (5.8) extends to all mean zero Lipschitz functions. In particular, by Proposition 1.14 every 1-Lipschitz function F on $\mathbb{R}^n$ is integrable with respect to μ and, for every $r \geq 0$,

$$\mu\big(\{F \geq \textstyle\int F d\mu + r\}\big) \leq e^{-r^2/2C}.$$

Furthermore

$$\alpha_{(\mathbb{R}^n,\mu)}(r) \leq e^{-r^2/8C}, \quad r > 0.$$

The preceding argument actually extends to measures on arbitrary metric spaces provided a natural extension of the length of the gradient is chosen. As in Section 3.1, given a locally Lipschitz function f on a metric space (X, d), define the length of the gradient of f at the point $x \in X$ as

$$|\nabla f|(x) = \limsup_{y \to x} \frac{|f(x) - f(y)|}{d(x,y)}. \tag{5.9}$$

This gradient satisfies the chain rule

$$|\nabla\phi(f)| \leq |\phi'(f)||\nabla f| \tag{5.10}$$

for $\phi : \mathbb{R} \to \mathbb{R}$ smooth enough. Note that $|\nabla f|(x) \leq \|F\|_{\mathrm{Lip}}$ at any x. By Rademacher's theorem, a Lipschitz function F on $\mathbb{R}^n$ is almost everywhere differentiable and $\|\nabla F\|_\infty = \|F\|_{\mathrm{Lip}}$. Furthermore, on $\mathbb{R}^n$, (5.1) extends to all locally Lipschitz functions. With a proof that just repeats the case of $\mathbb{R}^n$, the next result indicates that under a logarithmic Sobolev inequality, the underlying measure has normal concentration (Herbst's argument).

Theorem 5.3. *Let μ be a probability measure on the Borel sets of a metric space (X, d) such that for some $C > 0$ and all locally Lipschitz functions f on X,*

$$\mathrm{Ent}_\mu(f^2) \leq 2C \int |\nabla f|^2 d\mu.$$

Then, every 1-Lipschitz function $F : X \to \mathbb{R}$ is integrable and such that for every $r \geq 0$,

$$\mu\big(\{F \geq \textstyle\int F d\mu + r\}\big) \leq e^{-r^2/2C}.$$

In particular,

$$\alpha_{(X,d,\mu)}(r) \leq \mathrm{e}^{-r^2/8C}, \quad r > 0.$$

The normal concentration produced by Theorem 5.3 is optimal as shown by the example of the Gaussian measure γ for which $C = 1$. We saw in Corollary 3.2 that if a probability measure μ on (X, d) satisfies the Poincaré inequality

$$\mathrm{Var}_\mu(f) \leq C \int |\nabla f|^2 d\mu$$

for all locally Lipschitz functions f on (X, d), then μ has exponential concentration. Theorem 5.3 thus improves this result to normal concentration under a logarithmic Sobolev inequality. Therefore, as spectrum was used in Section 3.1 to recover the exponential isoperimetric concentration of Proposition 2.12, entropy and logarithmic Sobolev inequalities may be used to reach the Gaussian isoperimetric concentration of Theorem 2.7. As for Poincaré inequalities, note that if μ on (X, d) satisfies the logarithmic Sobolev inequality of Theorem 5.3 with constant C, the pushed forward measure μ_φ by a 1-Lipschitz map $\varphi : (X, d) \to (Y, \delta)$ also satisfies the same inequality. On a Riemannian manifold, Theorem 5.3 yields the following consequence that must be compared with Theorem 3.1 and that may be used to produce further examples of normal Lévy families in the sense of Section 3.3. Recall the logarithmic Sobolev constant ρ_0 of (5.5).

Corollary 5.4. *Let (X, g) be a compact Riemannian manifold with normalized Riemannian measure μ. Then,*

$$\alpha_{(X,g,\mu)}(r) \leq \mathrm{e}^{-\rho_0 r^2/8}, \quad r > 0,$$

where $\rho_0 > 0$ is the logarithmic Sobolev constant of the Laplace operator Δ on (X, g).

By (5.6), Corollary 5.4 covers Theorem 2.4. It might be worthwhile noting that Corollary 5.4 together with (5.7) improves the numerical constant of Theorem 2.3 including the one-dimensional torus.

To conclude this section, we briefly mention that both Poincaré and logarithmic Sobolev inequalities are stable under perturbation by a bounded potential. This is the content of the next simple proposition. One odd feature of this result is that it usually yields rather poor constants as functions of the dimension.

Proposition 5.5. *Let μ be a probability measure on the Borel sets of $\mathbb{R}^n$ and let V be bounded on $\mathbb{R}^n$. Define $d\nu = Z^{-1}\mathrm{e}^V d\mu$ where Z is the normalization factor. Then, if μ satisfies a Poincaré or logarithmic Sobolev inequality with constant C, then ν satisfies the same inequality with the constant $C\,\mathrm{e}^{4\|V\|_\infty}$.*

Proof. Note first that $\mathrm{e}^{-\|V\|_\infty} \leq Z \leq \mathrm{e}^{\|V\|_\infty}$. Let ζ be a smooth convex function on some open interval I of the real line. Typically, $\zeta(u) = u^2$ on $\mathbb{R}$ or $\zeta(u) = u \log u$ on $\mathbb{R}_+$. By convexity of ζ, for any probability measure μ and any smooth (bounded) $f : I \to \mathbb{R}$,

$$\int \zeta(f) d\mu - \zeta\Big(\int f d\mu\Big) = \inf_{t \in I} \int \big[\zeta(f) - \zeta(t) + (t - f)\zeta'(t)\big] d\mu \tag{5.11}$$

where the integrand on the right-hand side is non-negative. It immediately follows that

$$\int \zeta(f)d\nu - \zeta\Big(\int f d\nu\Big) \leq e^{2\|V\|_\infty}\Big[\int \zeta(f)d\mu - \zeta\Big(\int f d\mu\Big)\Big].$$

Hence, if

$$\mathrm{Var}_\mu(f) \leq C\int |\nabla f|^2 d\mu,$$

the choice of $\zeta(u) = u^2$ yields

$$\mathrm{Var}_\nu(f) \leq Ce^{2\|V\|_\infty}\int |\nabla f|^2 d\mu \leq Ce^{4\|V\|_\infty}\int |\nabla f|^2 d\nu.$$

If

$$\mathrm{Ent}_\mu(f^2) \leq 2C\int |\nabla f|^2 d\mu,$$

then, taking $\zeta(u) = u\log u$,

$$\mathrm{Ent}_\nu(f^2) \leq 2Ce^{2\|V\|_\infty}\int |\nabla f|^2 d\mu \leq 2Ce^{4\|V\|_\infty}\int |\nabla f|^2 d\nu.$$

Proposition 5.5 is proved. □

5.2 Product measures

One important feature of entropy (and of variance) is its product property. Together with the Herbst argument, this will give rise to a powerful tool to analyze dimension free concentration properties of product measures $P = \mu_1 \otimes \cdots \otimes \mu_n$. To illustrate the usefulness of this tool in product spaces, recall from Chapter 1 that the ℓ^1-metric cannot reflect dimension free concentration properties since the Lipschitz norm is not additive. We thus have to work at a higher level that motivates the interest for logarithmic Sobolev inequalities that deal (on $\mathbb{R}^n$ for example) with the energy

$$\int |\nabla f|^2 dP = \int \sum_{i=1}^n (\partial_i f)^2 d\mu_1 \otimes \cdots \otimes d\mu_n$$

which is clearly better adapted to product spaces than the Lipschitz coefficient $\|f\|_{\mathrm{Lip}} = \|\nabla f\|_\infty$.

Assume thus we are given probability spaces $(\Omega_i, \Sigma_i, \mu_i)$, $i = 1, \ldots, n$. Denote by P the product probability measure $\mu_1 \otimes \cdots \otimes \mu_n$ on the product space $X = \Omega_1 \times \cdots \times \Omega_n$ equipped with the product σ-field. A point x in X is denoted $x = (x_1, \ldots, x_n)$, $x_i \in \Omega_i$, $i = 1, \ldots, n$. Given f on the product space, we write furthermore f_i, $i = 1, \ldots, n$, for the function on Ω_i defined by

$$f_i(x_i) = f(x_1, \ldots, x_{i-1}, x_i, x_{i+1}, \ldots, x_n),$$

with $x_1, \ldots, x_{i-1}, x_{i+1}, \ldots, x_n$ fixed. The first result is the additivity of entropy.

Proposition 5.6. *For every non-negative function f on the product space X,*

$$\mathrm{Ent}_P(f) \leq \sum_{i=1}^n \int \mathrm{Ent}_{\mu_i}(f_i)dP.$$

The statement is the same for variances, namely for any f on the product space,

$$\mathrm{Var}_P(f) \leq \sum_{i=1}^n \int \mathrm{Var}_{\mu_i}(f_i)dP. \tag{5.12}$$

Proof. We only prove it for entropy, the proof for variance being entirely similar, and even simpler. For a non-negative function f on some probability space (Ω, Σ, μ),

$$\mathrm{Ent}_\mu(f) = \sup\left\{ \int fg d\mu; \int \mathrm{e}^g d\mu \leq 1 \right\}. \tag{5.13}$$

Indeed, assume by homogeneity that $\int f d\mu = 1$. By Young's inequality

$$uv \leq u \log u - u + \mathrm{e}^v, \quad u \geq 0, \quad v \in \mathbb{R},$$

we get, for $\int \mathrm{e}^g d\mu \leq 1$,

$$\int fg d\mu \leq \int f \log f d\mu - 1 + \int \mathrm{e}^g d\mu \leq \int f \log f d\mu.$$

The converse is obvious.

To prove Proposition 5.6, given g such that $\int \mathrm{e}^g dP \leq 1$, set, for every $i = 1, \dots, n$,

$$g^i(x_i, \dots, x_n) = \log\left(\frac{\int \mathrm{e}^{g(x_1,\dots,x_n)} d\mu_1(x_1) \cdots d\mu_{i-1}(x_{i-1})}{\int \mathrm{e}^{g(x_1,\dots,x_n)} d\mu_1(x_1) \cdots d\mu_i(x_i)} \right).$$

Then $g \leq \sum_{i=1}^n g^i$ and $\int \mathrm{e}^{(g^i)_i} d\mu_i = 1$. Therefore,

$$\begin{aligned}
\int fg dP &\leq \sum_{i=1}^n \int fg^i dP \\
&= \sum_{i=1}^n \int \left(\int f_i (g^i)_i d\mu_i \right) dP \\
&\leq \sum_{i=1}^n \int \mathrm{Ent}_{\mu_i}(f_i) dP
\end{aligned}$$

which is the result. Proposition 5.6 is established. □

Corollary 5.7. *Let μ_i on (X_i, d_i) satisfy the logarithmic Sobolev inequality*

$$\mathrm{Ent}_{\mu_i}(f^2) \leq 2C_i \int |\nabla_i f|^2 d\mu_i$$

for every locally Lipschitz function f on X_i, $i = 1, \ldots, n$, where $|\nabla_i f|$ is the generalized modulus of gradient in the i-th space X_i. Then the product measure $P = \mu_1 \otimes \cdots \otimes \mu_n$ on the product space $X = X_1 \times \cdots \times X_n$ satisfies the logarithmic Sobolev inequality

$$\mathrm{Ent}_P(f^2) \leq 2 \max_{1 \leq i \leq n} C_i \int |\nabla f|^2 dP$$

for every locally Lipschitz function f on X where

$$|\nabla f|^2 = \sum_{i=1}^{n} |\nabla_i f|^2.$$

The same assertion holds for the Poincaré inequality. In particular, if (X_i, g_i), $i = 1, \ldots, n$, are compact Riemannian manifolds, and if X is the Riemannian product of the X_i's,

$$\lambda_1(X) = \min_{1 \leq i \leq n} \lambda_1(X_i) \quad \text{and} \quad \rho_0(X) = \min_{1 \leq i \leq n} \rho_0(X_i).$$

The main consequence of Corollary 5.7 is that it yields, together with the Herbst argument, concentration results in product spaces with respect to the Euclidean metric $\big(\sum_{i=1}^n d_i^2\big)^{1/2}$ which are independent of the dimension. Corollary 5.7 may be combined in applications with the perturbation result of Proposition 5.5.

The following is a simple consequence of Proposition 5.6 that bounds in a useful way the entropies along each coordinate.

Corollary 5.8. *For every function f on the product space $X = \Omega_1 \times \cdots \times \Omega_n$ and every product probability measure $P = \mu_1 \otimes \cdots \otimes \mu_n$,*

$$\mathrm{Ent}_P(\mathrm{e}^f) \leq \frac{1}{2} \sum_{i=1}^{n} \int \mathcal{R}_i(\mathrm{e}^{f_i})(x) dP(x)$$

where, for $i = 1, \ldots, n$,

$$\mathcal{R}_i(\mathrm{e}^{f_i})(x) = \iint_{\{f_i(x_i) \geq f_i(y_i)\}} \big[f_i(x_i) - f_i(y_i)\big]^2 \mathrm{e}^{f_i(x_i)} d\mu_i(x_i) d\mu_i(y_i).$$

Proof. The proof is elementary. We may assume f bounded. By the product property of entropy (Proposition 5.6), it is enough to deal with the case $n = 1$. By Jensen's inequality,

$$\mathrm{Ent}_P(\mathrm{e}^f) \leq \int f \mathrm{e}^f dP - \int \mathrm{e}^f dP \int f dP.$$

The right-hand side of the latter may then be rewritten as

$$\begin{aligned} &\frac{1}{2} \iint \big[f(x) - f(y)\big] \big[\mathrm{e}^{f(x)} - \mathrm{e}^{f(y)}\big] dP(x) dP(y) \\ &\quad = \iint_{\{f(x) \geq f(y)\}} \big[f(x) - f(y)\big] \big[\mathrm{e}^{f(x)} - \mathrm{e}^{f(y)}\big] dP(x) dP(y). \end{aligned}$$

Since for $u \geq v$

$$(u-v)(\mathrm{e}^u - \mathrm{e}^v) \leq \tfrac{1}{2}\,(u-v)^2(\mathrm{e}^u + \mathrm{e}^v) \leq (u-v)^2\mathrm{e}^u,$$

the conclusion easily follows. Corollary 5.8 is established. □

On the basis of these results, we first illustrate the use of logarithmic Sobolev inequalities in the basic example of the Hamming metric. As a consequence of Corollary 5.8, if

$$N(f) = \sup_{x\in X}\left(\sum_{i=1}^n \int \left|f(x) - f_i(y_i)\right|^2 d\mu_i(y_i)\right)^{1/2},$$

then

$$\mathrm{Ent}_P(\mathrm{e}^f) \leq N(f)^2 \int \mathrm{e}^f\, dP. \tag{5.14}$$

If F is a 1-Lipschitz function with respect to the Hamming metric on the product space X then $N(F)^2 \leq n$. Now, the proof of Theorem 5.3 above immediately shows from (5.14) that if F is mean zero and 1-Lipschitz, then for every $\lambda \in \mathbb{R}$,

$$\int \mathrm{e}^{\lambda F} dP \leq \mathrm{e}^{n\lambda^2}.$$

Therefore, by Proposition 1.14, for any product probability P on the product space $X = \Omega_1 \times \cdots \times \Omega_n$ equipped with the Hamming metric,

$$\alpha_{(X,d,P)}(r) \leq \mathrm{e}^{-r^2/16n}, \quad r > 0,$$

that should be compared once more to (1.24) (and (4.2)).

In the last part of this section, we provide along the same lines an elementary approach to Corollary 4.10. The point is that while the deviation inequalities have no reason to be tensorizable, they are actually consequences of a logarithmic Sobolev inequality, which only needs to be proved in dimension one. The main result in this direction is the following statement. Let $\mu_1, \ldots, \mu_n$ be arbitrary probability measures on the unit interval $[0,1]$ and let P be the product probability measure $P = \mu_1 \otimes \cdots \otimes \mu_n$ on $[0,1]^n$. We say that a function on $\mathbb{R}^n$ is separately convex if it is convex in each coordinate. Recall that a convex function on $\mathbb{R}$ is continuous and almost everywhere differentiable.

Theorem 5.9. *Let F be separately convex and 1-Lipschitz on $\mathbb{R}^n$. Then, for every product probability measure P on $[0,1]^n$, and every $r \geq 0$,*

$$P\big(\{F \geq \textstyle\int F dP + r\}\big) \leq \mathrm{e}^{-r^2/4}.$$

Proof. The theorem will follow from the following stronger statement of independent interest.

Proposition 5.10. *For every separately convex function f and every product probability P on $\mathbb{R}^n$,*

$$\mathrm{Ent}_P(\mathrm{e}^f) \leq \iint \sum_{i=1}^n (x_i - y_i)^2 (\partial_i f)(x)^2\, \mathrm{e}^{f(x)} dP(x)dP(y).$$

Proof. By Corollary 5.8, it is enough to show that for every $i = 1, \ldots, n$ and every $x = (x_1, \ldots, x_n) \in \mathbb{R}^n$,

$$\mathcal{R}_i(\mathrm{e}^{f_i})(x) \leq \iint (x_i - y_i)^2 {f_i}'(x_i)^2 \, \mathrm{e}^{f_i(x_i)} d\mu_i(x_i) d\mu_i(y_i).$$

Now, since f is separately convex, for all $x, y \in \mathbb{R}^n$,

$$f_i(x_i) - f_i(y_i) \leq (x_i - y_i) f_i'(x_i).$$

The proof is easily completed. Proposition 5.10 is established. □

We now complete the proof of Theorem 5.9 following the Herbst argument. Let F be a separately convex Lipschitz function such that $\|F\|_{\mathrm{Lip}} \leq 1$ by homogeneity. By Rademacher's theorem, $|\nabla F| \leq 1$ almost everywhere. Replacing F by a convolution with a Gaussian kernel, we may actually suppose that $|\nabla F| \leq 1$ everywhere. Then, the argument is entirely similar to the proof of Theorem 5.3. Applying indeed Proposition 5.10 to $\lambda F - \lambda^2$, $\lambda \geq 0$, we deduce that

$$\mathrm{Ent}_P(\mathrm{e}^{\lambda F - \lambda^2}) \leq \lambda^2 \int |\nabla F|^2 \mathrm{e}^{\lambda F - \lambda^2} dP$$

where we used that P is concentrated on $[0,1]^n$. Therefore, for every $\lambda \geq 0$,

$$\lambda \Lambda'(\lambda) \leq \Lambda(\lambda) \log \Lambda(\lambda)$$

where $\Lambda(\lambda) = \int \mathrm{e}^{\lambda F - \lambda^2} dP$. The proof is completed as in Theorem 5.3. □

It should be pointed out that Theorem 5.9 does not produce deviation inequalities under the mean since we have to work with $\lambda \geq 0$ to preserve convexity. The point is that Theorem 5.9 is stated for separately convex functions, whereas deviations under the mean seem to require the full convexity assumption. (See [Sa] for a discussion of deviation inequalities under the mean of convex functions with respect to product measures.)

Proposition 5.10 is of particular use for norms of sums of independent random vectors and will be crucial in the investigation of sharp bounds in Section 7.3. The proposition also puts forward the generalized gradient (in dimension one)

$$|\nabla f|(x) = \left(\int (x - y)^2 f'(y)^2 d\mu(y) \right)^{1/2}$$

of statistical interest.

5.3 Modified logarithmic Sobolev inequalities

As we have seen in Section 4.5, products of the usual exponential distribution somewhat surprisingly satisfy a concentration property which, in some respect, is stronger than concentration for Gaussian measures. Our first aim here will be to show that this result can be seen as a consequence of some appropriate logarithmic Sobolev inequality which we call modified. One of its main interest is that it tensorizes with two parameters on the gradient, one on its supremum norm and one on the usual quadratic norm. This feature is the appropriate explanation for the concentration property of the exponential distribution.

We first investigate a modified logarithmic Sobolev inequality for the exponential distribution. We then describe the product properties of modified logarithmic Sobolev inequalities and their applications to concentration. We then observe that all measures satisfying a Poincaré inequality do satisfy the same modified inequality as the exponential distribution.

Let us start back with the concentration property described by Theorem 4.16. Let ν^n be the product measure on $\mathbb{R}^n$ when each factor is endowed with the measure ν of density $\frac{1}{2}\,\mathrm{e}^{-|x|}$ with respect to Lebesgue measure. Recall $\mathcal{B}_2^n$ the Euclidean unit ball and $\mathcal{B}_1^n$ the ℓ^1-unit ball in $\mathbb{R}^n$. Then, for every Borel set A in $\mathbb{R}^n$ and every $r > 0$,

$$1 - \nu^n\big(A + 6\sqrt{r}\,\mathcal{B}_2^n + 9r\mathcal{B}_1^n\big) \le \frac{1}{\nu^n(A)}\,\mathrm{e}^{-r}. \tag{5.15}$$

As we have seen in Proposition 4.18, this concentration property on sets may be translated equivalently on functions. Let F be a real-valued function on $\mathbb{R}^n$ such that $\|F\|_{\mathrm{Lip}} \le a$ and with Lipschitz coefficient b with respect to the ℓ^1-metric, that is,

$$\big|F(x) - F(y)\big| \le b\sum_{i=1}^n |x_i - y_i|, \quad x, y \in \mathbb{R}^n.$$

Then, for every $r \ge 0$,

$$\nu^n\big(\{F \ge M + r\}\big) \le \exp\bigg(-\frac{1}{K}\,\min\bigg(\frac{r}{b}, \frac{r^2}{a^2}\bigg)\bigg) \tag{5.16}$$

for some numerical constant $K > 0$ where M is a median or the mean of F for ν_n.

Our first task will be to present an elementary proof of (5.16) based on logarithmic Sobolev inequalities. Following the Herbst argument in the case of the exponential distribution would require to determine the appropriate logarithmic Sobolev inequality satisfied by ν^n. We cannot hope for a classical logarithmic Sobolev inequality to hold simply because it would imply that linear functions have a Gaussian tail for ν^n. To investigate logarithmic Sobolev inequalities for ν^n, it is enough, by the fundamental product property of entropy, to deal with the dimension one. One first inequality may be deduced from the Gaussian logarithmic Sobolev inequality. Given a smooth function f on $\mathbb{R}$, apply Theorem 5.1 in dimension 2 to $g(x,y) = f(\frac{x^2+y^2}{2})$. Letting $\tilde{\nu}$ denote the one-sided exponential distribution with density e^{-x} with respect to Lebesgue measure on $\mathbb{R}_+$, we get

$$\mathrm{Ent}_{\tilde{\nu}}(f^2) \le 4\int x f'(x)^2 d\tilde{\nu}(x).$$

If $\tilde{\nu}^n$ denote the product measure on $\mathbb{R}_+^n$, we have similarly for every smooth f on $\mathbb{R}_+^n$,

$$\mathrm{Ent}_{\tilde{\nu}^n}(f^2) \le 4\int \sum_{i=1}^n x_i\big|\partial_i f(x)\big|^2 d\tilde{\nu}^n(x). \tag{5.17}$$

It does not seem however that this logarithmic Sobolev inequality (5.17) can yield (5.16), and thus the geometric concentration (5.15), via the Herbst argument. In a sense, this negative observation is compatible with the fact that (5.15) improves upon some aspects of the concentration for Gaussian measures as we have seen in Section 4.5. We thus have to look for some other version of the logarithmic Sobolev inequality for the exponential distribution. To this task, let us observe that, at the level of Poincaré inequalities, there are two distinct inequalities. For simplicity, let us deal again only with $n = 1$. The first one, in the spirit of (5.17), is

$$\mathrm{Var}_{\tilde{\nu}}(f) \leq \int x f'(x)^2 d\tilde{\nu}(x). \tag{5.18}$$

This may be shown, either from the Gaussian Poincaré inequality as before, with however a worse constant, or by noting that the first eigenvalue of the Laguerre generator with invariant measure $\tilde{\nu}$ is 1. Actually, (5.17) is exactly the (optimal) logarithmic Sobolev inequality associated to the Laguerre generator. By the way, that 4 is the best constant in (5.17) is an easy consequence of our arguments. Indeed, if (5.17) holds with a constant $C < 4$, a function F on $\mathbb{R}_+$ such that $xF'(x)^2 \leq 1$ almost everywhere would be such that $\int e^{F^2/4} d\tilde{\nu} < \infty$ by Theorem 5.3 (together with Proposition 1.9). But the example of $F(x) = 2\sqrt{x}$ contradicts this consequence (cf. [Ko-S]).

A second Poincaré inequality states that

$$\mathrm{Var}_{\tilde{\nu}}(f) \leq 4 \int f'^2 d\tilde{\nu}. \tag{5.19}$$

Inequalities (5.18) and (5.19) are not directly comparable and, in a sense, we are looking for an analogue of (5.19) for entropy.

To introduce this result, let us first recall the proof of (5.19). We will work with the double exponential distribution ν. It is plain that all the results hold, with the obvious modifications, for the one-sided exponential distribution $\tilde{\nu}$. We also work below with smooth functions. (One may consider for example the space of all continuous almost everywhere differentiable functions $f : \mathbb{R}^n \to \mathbb{R}$ such that $\int |f| d\nu^n < \infty$, $\int |\nabla f| d\nu^n < \infty$ and

$$\lim_{x_i \to \pm\infty} e^{-|x_i|} f(x_1, \ldots, x_i, \ldots, x_n) = 0$$

for every $i = 1, \ldots, n$ and $x_1, \ldots, x_{i-1}, x_{i+1}, \ldots, x_n \in \mathbb{R}$.) The main argument of the proof is the following simple observation. If ϕ is smooth on $\mathbb{R}$, by the integration by parts formula,

$$\int \phi d\nu = \phi(0) + \int \mathrm{sgn}(x) \phi'(x) d\nu(x). \tag{5.20}$$

Lemma 5.11. *For every smooth f on $\mathbb{R}$,*

$$\mathrm{Var}_\nu(f) \leq 4 \int f'^2 d\nu.$$

Proof. Set $g(x) = f(x) - f(0)$. Then, by (5.20) and the Cauchy-Schwarz inequality,

$$\begin{aligned}\int g^2 d\nu &= 2 \int \mathrm{sgn}(x) g'(x) g(x) d\nu(x) \\ &\leq 2 \Big(\int g'^2 d\nu \Big)^{1/2} \Big(\int g^2 d\nu \Big)^{1/2}.\end{aligned}$$

Since $\mathrm{Var}_\nu(f) = \mathrm{Var}_\nu(g) \le \int g^2 d\nu$, and $g' = f'$, the lemma follows. □

We turn to the corresponding inequality for entropy and the main result of this section.

Theorem 5.12. *For every $0 < \rho < 1$ and every Lipschitz function f on $\mathbb{R}$ such that $|f'| \le \rho < 1$ almost everywhere,*

$$\mathrm{Ent}_\nu(\mathrm{e}^f) \le \frac{2}{1-\rho} \int f'^2 \mathrm{e}^f d\nu.$$

Note that Theorem 5.12, when applied to functions εf as $\varepsilon \to 0$, implies Lemma 5.11. Theorem 5.12 is the first example of what we will call a modified logarithmic Sobolev inequality. Theorem 5.12 is basically only used below for some fixed values of ρ, for example $\rho = \frac{1}{2}$.

Proof. Changing f into f+const we may assume that $f(0) = 0$. Since $u \log u \ge u - 1$, $u \ge 0$, we have

$$\mathrm{Ent}_\nu(\mathrm{e}^f) \le \int \left[f\mathrm{e}^f - \mathrm{e}^f + 1\right] d\nu.$$

Since $|f'| \le \rho < 1$ almost everywhere, the functions e^f, $f\mathrm{e}^f$ and $f^2\mathrm{e}^f$ are smooth in our preceding sense. Therefore, by repeated use of (5.20),

$$\int \left[f\mathrm{e}^f - \mathrm{e}^f + 1\right] d\nu = \int \mathrm{sgn}(x) f'(x) f(x)\, \mathrm{e}^{f(x)} d\nu(x)$$

and

$$\begin{aligned}\int f^2 \mathrm{e}^f d\nu &= 2\int \mathrm{sgn}(x) f'(x) f(x)\, \mathrm{e}^{f(x)} d\nu(x)\\ &\quad + \int \mathrm{sgn}(x) f'(x) f(x)^2 \mathrm{e}^{f(x)} d\nu(x).\end{aligned}$$

By the Cauchy-Schwarz inequality and the assumption on f',

$$\int f^2 \mathrm{e}^f d\nu \le 2\left(\int f'^2 \mathrm{e}^f d\nu\right)^{1/2} \left(\int f^2 \mathrm{e}^f d\nu\right)^{1/2} + \rho \int f^2 \mathrm{e}^f d\nu$$

so that

$$\int f^2 \mathrm{e}^f d\nu \le \left(\frac{2}{1-\rho}\right)^2 \int f'^2 \mathrm{e}^f d\nu.$$

Now, by the Cauchy-Schwarz inequality again,

$$\begin{aligned}\mathrm{Ent}_\nu(\mathrm{e}^f) &\le \int \mathrm{sgn}(x) f'(x) f(x)\, \mathrm{e}^{f(x)} d\nu(x)\\ &\le \left(\int f'^2 \mathrm{e}^f d\nu\right)^{1/2} \left(\int f^2 \mathrm{e}^f d\nu\right)^{1/2}\\ &\le \frac{2}{1-\rho} \int f'^2 \mathrm{e}^f d\nu\end{aligned}$$

which is the result. Theorem 5.12 is proved. □

We are now ready to describe the application to concentration. As a consequence of Theorem 5.12 and of the product property of entropy (Proposition 5.6), for every smooth enough function F on $\mathbb{R}^n$ such that $\max_{1\le i\le n}|\partial_i F| \le 1$ almost everywhere and every λ, $|\lambda| \le \rho < 1$,

$$\mathrm{Ent}_{\nu^n}(\mathrm{e}^{\lambda F}) \le \frac{2\lambda^2}{1-\rho}\int\sum_{i=1}^n(\partial_i F)^2\mathrm{e}^{\lambda F}d\nu. \tag{5.21}$$

For simplicity, let us take $\rho = \frac{1}{2}$ (although $\rho < 1$ might improve some numerical constants below). Moreover, assume that $\sum_{i=1}^n(\partial_i F)^2 \le a^2$ almost everywhere. Then, by (5.21),

$$\mathrm{Ent}_{\nu^n}(\mathrm{e}^{\lambda F}) \le 4a^2\lambda^2\int \mathrm{e}^{\lambda F}d\nu^n$$

for every $|\lambda| \le \frac{1}{2}$. We now argue as in the proof of Theorem 5.3. Let $\Lambda(\lambda) = \int \mathrm{e}^{\lambda F - 4a^2\lambda^2}d\nu^n$, $|\lambda| \le \frac{1}{2}$. The preceding inequality expresses that

$$\lambda\Lambda'(\lambda) \le \Lambda(\lambda)\log\Lambda(\lambda), \quad |\lambda| \le \tfrac{1}{2}\,.$$

Integrating this differential inequality shows that

$$\int \mathrm{e}^{\lambda F}d\nu^n \le \mathrm{e}^{\int F d\nu^n + 4a^2\lambda^2},$$

which only holds for $|\lambda| \le \frac{1}{2}$. By Chebyshev's exponential inequality, for every $r \ge 0$ and $|\lambda| \le \frac{1}{2}$,

$$\nu^n\big(\{F \ge \textstyle\int F d\nu^n + r\}\big) \le \mathrm{e}^{-\lambda r + 4a^2\lambda^2}.$$

Minimizing over λ (take $\lambda = r/8a^2$ if $r \le 4a^2$ and $\lambda = \frac{1}{2}$ if $r \ge 4a^2$) shows that, for every $r \ge 0$,

$$\nu^n\big(\{F \ge \textstyle\int F d\nu^n + r\}\big) \le \exp\bigg(-\frac{1}{4}\min\Big(r, \frac{r^2}{4a^2}\Big)\bigg).$$

By homogeneity, this inequality amounts to (5.16) (with $K = 16$) and our claim is proved. As already mentioned, we have a similar result for the one-sided exponential distribution.

The inequality put forward in Theorem 5.12 for the exponential distribution is an example of what can be called modified logarithmic Sobolev inequalities. In order to describe this notion in some generality, we set the following definition. It allows us to describe various types of concentration behavior by the appropriate functional logarithmic Sobolev inequality. We take again the framework of a metric space (X, d) with the generalized length of a gradient $|\nabla f|$. We say that a probability measure μ on the Borel sets of a metric space (X, d) satisfies a *modified logarithmic Sobolev inequality* if there is a function $\beta(\rho) \ge 0$ on $\mathbb{R}_+$ such that, whenever $\|\nabla f\|_\infty \le \rho$,

$$\mathrm{Ent}_\mu(\mathrm{e}^f) \le \beta(\rho)\int |\nabla f|^2\,\mathrm{e}^f d\mu \tag{5.22}$$

for all f's such that $\int \mathrm{e}^f d\mu < \infty$.

According to Theorem 5.12, the exponential distribution ν on the line satisfies a modified logarithmic Sobolev inequality with respect to the usual gradient with $\beta(\rho)$ bounded for the small values of ρ. On the other hand, the canonical Gaussian measure γ satisfies a modified logarithmic Sobolev inequality with $\beta(\rho) = \frac{1}{2}$, $\rho \geq 0$.

The definition (5.22) implies that

$$\mathrm{Ent}_\mu(\mathrm{e}^f) \leq \rho^2 \beta(\rho) \int \mathrm{e}^f \, d\mu$$

for every f with $\|\nabla f\|_\infty \leq \rho$. In particular, if $\beta(\rho)$ is bounded for the small values of ρ, Lipschitz functions will have an exponential tail.

The main new feature here is that the modified logarithmic Sobolev inequalities tensorize in terms of two parameters rather than only the Lipschitz bound. This property is summarized in the next proposition which is an elementary consequence of the product property of entropy (Proposition 5.6). It allows new concentration behaviors.

Let $\mu_1, \ldots, \mu_n$ be probability measures on respective metric spaces (X_i, d_i), $i = 1, \ldots, n$, and let $P = \mu_1 \otimes \cdots \otimes \mu_n$ be the product probability on the product space $X = X_1 \times \cdots \times X_n$. If f is a function on the product space, for each i, f_i is the function f depending on the i-th variable with the other coordinates fixed. As in Corollary 5.7, we denote by $|\nabla_i f|$ the generalized modulus of gradient in the i-th space X_i.

Proposition 5.13. *Assume that for every function f on X_i such that $\|\nabla_i f\|_\infty \leq \rho$,*

$$\mathrm{Ent}_{\mu_i}(\mathrm{e}^f) \leq \beta(\rho) \int |\nabla_i f|^2 \mathrm{e}^f \, d\mu_i, \quad i = 1, \ldots, n.$$

Then, for every f on the product space such that $\max_{1 \leq i \leq n} \|\nabla_i f_i\|_\infty \leq \rho$,

$$\mathrm{Ent}_P(\mathrm{e}^f) \leq \beta(\rho) \int |\nabla f|^2 \mathrm{e}^f \, dP$$

where $|\nabla f|^2 = \sum_{i=1}^n |\nabla_i f|^2$.

According to the behavior of $\beta(\rho)$, this proposition yields concentration properties in terms of the parameters

$$\Big\| \sum_{i=1}^n |\nabla_i f|^2 \Big\|_\infty \quad \text{and} \quad \max_{1 \leq i \leq n} \|\nabla_i f\|_\infty .$$

For example, if $\beta(\rho) \leq c$ for $0 \leq \rho \leq \rho_0$, the product measure P will satisfy the same concentration inequality as the one for the exponential distribution (5.16). In the next section, we investigate instances where $\beta(\rho) \leq \kappa \, \mathrm{e}^{\delta\rho}$, $\rho \geq 0$, related to the Poisson measure.

We now observe that the concentration properties of the exponential distribution are actually shared by all measures satisfying a Poincaré inequality. More precisely, every such measure satisfies the modified logarithmic Sobolev inequality of Theorem 5.12.

Thus let $|\nabla f|$ be a generalized modulus of gradient (5.9) on a metric space (X, d), satisfying thus the chain rule formula (5.10). Let μ be a probability measure on the Borel σ-field of X such that for some $C > 0$ and all locally Lipschitz functions f,

$$\mathrm{Var}_\mu(f) \leq C \int |\nabla f|^2 d\mu. \tag{5.23}$$

We already know from Corollary 3.2 that such a Poincaré inequality implies exponential integrability of Lipschitz functions. We next show how it also implies a modified logarithmic Sobolev inequality which yields, as just discussed, concentration properties for the product measures μ^n similar to those of the products of the exponential distribution. We refer to [Bo-L1] for a proof of Theorem 5.14 below that extends the argument of Theorem 5.12.

Theorem 5.14. *Let μ be a probability measure on the Borel sets of a metric space (X, d) satisfying the Poincaré inequality (5.23). Then, for any function f on X such that $\|\nabla f\|_\infty \leq \rho < 2/\sqrt{C}$,*

$$\mathrm{Ent}_\mu(\mathrm{e}^f) \leq \beta(\rho) \int |\nabla f|^2 \mathrm{e}^f d\mu$$

where

$$\beta(\rho) = \frac{C}{2} \left(\frac{2 + \rho\sqrt{C}}{2 - \rho\sqrt{C}} \right)^2 \mathrm{e}^{\sqrt{5C}\rho}.$$

Note that $\beta(\rho)$ is uniformly bounded for the small values of ρ, for example $\beta(\rho) \leq 3\,\mathrm{e}^5 C/2$ when $\rho \leq 1/\sqrt{C}$. As a corollary, we obtain, following the preceding discussion, a concentration inequality of exponential type for the product measure μ^n of μ on X^n.

Corollary 5.15. *Let μ satisfy (5.23) and denote by μ^n the n-fold product of μ on X^n. Then, every function F on X^n such that*

$$\sum_{i=1}^n |\nabla_i F|^2 \leq a^2 \quad \text{and} \quad \max_{1 \leq i \leq n} |\nabla_i F| \leq b$$

μ^n-almost everywhere is integrable with respect to μ^n, and for every $r \geq 0$,

$$\mu^n\big(\{F \geq \textstyle\int F d\mu^n + r\}\big) \leq \exp\left(-\frac{1}{K} \min\left(\frac{r}{b}, \frac{r^2}{a^2} \right) \right)$$

where $K > 0$ only depends on the constant C in the Poincaré inequality (5.23).

One may obtain a similar statement for products of possibly different measures μ with a uniform lower bound on the constants in the Poincaré inequalities (5.23).

Corollary 5.15 may be turned into an inequality on sets such as (5.15). More precisely, if $\mu^n(A) \geq \frac{1}{2}$, for every $r \geq 0$ and some numerical constant $K > 0$,

$$\mu^n\big(\{F_A \geq r\}\big) \leq \mathrm{e}^{-r/K}$$

where for $x = (x_1, \dots, x_n) \in X^n$ and $A \subset X^n$,

$$F_A(x) = \inf_{a \in A} \sum_{i=1}^{n} \min\big(d(x_i, a_i), d(x_i, a_i)^2\big).$$

An important feature of the constant $\beta(\rho)$ of Theorem 5.14 is that $\beta(\rho) \to C/2$ as $\rho \to 0$. In particular, the modified logarithmic Sobolev inequality of Theorem 5.14 implies back the Poincaré inequality (5.23) by applying it to functions εf with $\varepsilon \to 0$. The Poincaré inequality (5.23) and the modified logarithmic Sobolev inequality of Theorem 5.14 thus appear as formally equivalent.

Other exponential decays may also be produced by families of inequalities interpolating between Poincaré and logarithmic Sobolev inequalities considered recently by R. Latała and K. Oleskiewicz [L-O]. For a probability measure μ on the Borel sets of a metric space (X, d), consider the inequality, for some $\theta \in [0, 1]$ and $C > 0$, and every $1 \le q < 2$ and every locally Lipschitz function f on (X, d),

$$\int f^2 d\mu - \left(\int |f|^q d\mu\right)^{2/q} \le C(2-q)^\theta \int |\nabla f|^2 d\mu. \tag{5.24}$$

It is easily seen that these inequalities are stronger and stronger as θ increases and that $\theta = 0$ amounts to the Poincaré inequality whereas $\theta = 1$ amounts to the logarithmic Sobolev inequality (within numerical constants). One may also show (cf. [L-O]) that these families of inequalities for a given θ are stable under products. As a main result, it is shown in [L-O] that the measures ν_p with density $c_p\, \mathrm{e}^{-|x|^p}$, $1 \le p \le 2$, with respect to Lebesgue measure on $\mathbb{R}$ satisfy (5.24) with $\theta = 2(1 - \frac{1}{p})$. Together with the following, it may be used in particular to recover Proposition 4.21 (cf. [S-Z2]).

Theorem 5.16. *Let μ be a probability measure on the Borel sets of a metric space (X, d) satisfying (5.24) for some $\theta \in [0, 1]$ and $C > 0$. Then, every 1-Lipschitz function $F : X \to \mathbb{R}$ is integrable and such that for every $r \ge 0$,*

$$\mu\big(\{F \ge \textstyle\int F d\mu + r\}\big) \le \mathrm{e}^{-r^p/KC^{p/2}}$$

where $1 \le p \le 2$ is such that $\theta = 2(1 - \frac{1}{p})$ and where $K > 0$ is a numerical constant. In particular,

$$\alpha_{(X,d,\mu)}(r) \le \mathrm{e}^{-r^p/KC^{p/2}}, \quad r > 0.$$

5.4 Discrete settings

In this section, we briefly describe how the entropic method may be developed similarly in discrete structures for which the chain rule (5.10) on gradients is not available. We then discuss more simply the examples of uniform measure on the discrete cube $\{0, 1\}^n$ and of standard Poisson measure.

We take again the discrete setting of the end of Section 3.1 where exponential concentration for Markov chains and graphs under spectral properties was investigated. Recall that (Π, μ) is a reversible Markov chain with invariant measure μ on a finite or countable set X if $\Pi(x,y) \geq 0$ satisfies

$$\sum_{y \in X} \Pi(x,y) = 1$$

for every $x \in X$ and if $\Pi(x,y)\mu(\{x\})$ is symmetric in x and y and $\sum_x \Pi(x,y)\mu(\{x\}) = \mu(\{y\})$ for every $y \in X$. Define

$$\mathcal{Q}(f,f) = \frac{1}{2} \sum_{x,y \in X} \big[f(x) - f(y)\big]^2 \Pi(x,y)\mu(\{x\})$$

and set

$$|||f|||_\infty^2 = \frac{1}{2} \sup_{x \in X} \sum_{y \in X} \big|f(x) - f(y)\big|^2 \Pi(x,y).$$

We already mentioned that the $|||\cdot|||_\infty$-norm tries to be as close as possible to the sup-norm of a gradient in a continuous setting. The next statement is thus the analogue of Theorem 5.3 in this setting, and the logarithmic Sobolev version of the Poincaré Theorem 3.3.

Theorem 5.17. *Let (Π,μ) be a reversible Markov chain on X as before, and assume that for some constant $C > 0$ and all f's on X,*

$$\mathrm{Ent}_\mu(f^2) \leq 2C\mathcal{Q}(f,f).$$

Then, whenever $|||F|||_\infty \leq 1$, F is integrable with respect to μ and for every $r \geq 0$,

$$\mu\big(\{F \geq \textstyle\int F d\mu + r\}\big) \leq \mathrm{e}^{-r^2/4C}.$$

Proof. We apply the logarithmic Sobolev inequality to $f^2 = \mathrm{e}^{\lambda F}$, $\lambda \in \mathbb{R}$. Recall from (3.4) that

$$\mathcal{Q}\big(\mathrm{e}^{\lambda F/2}, \mathrm{e}^{\lambda F/2}\big) \leq |||F|||_\infty^2 \int \mathrm{e}^{\lambda F} d\mu.$$

Therefore, for every $\lambda \in \mathbb{R}$,

$$\mathrm{Ent}_\mu(\mathrm{e}^{\lambda F}) \leq C \int \mathrm{e}^{\lambda F} d\mu$$

from which the proof is completed as in Theorem 5.3. The result follows. □

While the norm $|||\cdot|||_\infty$ might appear as the proper generalization of the Lipschitz norm on X, it does not always reflect accurately discrete situations. Discrete gradients may also be examined in another way. For example, if f is a real-valued function on $\mathbb{Z}$, set

$$Df(x) = f(x+1) - f(x), \quad x \in \mathbb{Z}. \tag{5.25}$$

One may then regard as Lipschitz the norm $\sup_{x \in \mathbb{Z}^d} |Df(x)|$, which will prove more adapted to a number of cases, such as for example Poisson measures. The lack of chain rule (for example, $|D(\mathrm{e}^f)| \leq |Df|\mathrm{e}^{|Df|}\mathrm{e}^f$ only in general) will then have to be handled by other means, but will also determine the best that can be expected under some appropriate logarithmic Sobolev inequality. This is the subject to which we turn now. Actually, the norm $|||\cdot|||_\infty$ is just adapted to produce Gaussian bounds and, as such, is not suited to a number of various exponential decays.

The preceding example may be further generalized to $\mathbb{Z}^d$. Similarly, in the context of statistical mechanics, we may consider $X = \{-1,+1\}^{\mathbb{Z}^d}$ and let

$$|Df(\omega)| = \Big(\sum_{k\in\mathbb{Z}^d} |\partial_k f(\omega)|^2\Big)^{1/2}$$

where $\partial_k f(\omega) = f(\omega^k) - f(\omega)$ and where ω^k is the element of X obtained from ω by replacing the k-th coordinate with $-\omega_k$.

As a starting point of our investigation, consider the logarithmic Sobolev inequality for the uniform product measure $\mu = \mu^n$ on $\{0,1\}^n$ expressed by

$$\mathrm{Ent}_\mu(f^2) \le \frac{1}{2}\sum_{i=1}^n \int |D_i f|^2 d\mu \tag{5.26}$$

for any function f on $\{0,1\}^n$ where

$$D_i f(x) = f(x) - f\big(s_i(x)\big), \quad x = (x_1,\dots,x_n) \in \{0,1\}^n$$

with $s_i(x) = (x_1,\dots,x_{i-1},1-x_i,x_{i+1},\dots,x_n)$. The gradients D_i are just (5.25) on the two-point space. By the product property of entropy, this inequality needs only be proven in dimension one, for which it amounts to the inequality

$$u^2\log u^2 + v^2\log v^2 - (u^2+v^2)\log\Big(\frac{u^2+v^2}{2}\Big) \le (u-v)^2$$

for any real u, v. By means of the central limit theorem, it may be used to prove the logarithmic Sobolev inequality for Gaussian measures (cf. [Gros1]).

This logarithmic Sobolev inequality may be used to recover concentration with respect to the Hamming metric on $\{0,1\}^n$. Namely, let F be 1-Lipschitz with respect to the Hamming metric on the discrete cube $\{0,1\}^n$. Following the Herbst argument, apply (5.26) to $f^2 = \mathrm{e}^{\lambda F}$, $\lambda \in \mathbb{R}$. Although the D_i's do not satisfy the chain rule formula, since $|F(x) - F(s_i(x))| \le 1$, it is easily seen that

$$\big|D_i(\mathrm{e}^{\lambda F/2})(x)\big| = \big|\mathrm{e}^{\lambda F(x)/2} - \mathrm{e}^{\lambda F(s_i(x))/2}\big| \le |\lambda|\,\mathrm{e}^{\lambda F(x)/2}$$

for every $|\lambda| \le 1$. Hence, setting as usual $\Lambda(\lambda) = \int \mathrm{e}^{\lambda F - n\lambda^2/2} d\mu$, $\lambda \in \mathbb{R}$, we have

$$\lambda\Lambda'(\lambda) \le \Lambda(\lambda)\log\Lambda(\lambda), \quad |\lambda| \le 1.$$

We integrate this differential inequality as in Section 5.1 to get that

$$\int \mathrm{e}^{\lambda F} d\mu \le \mathrm{e}^{\lambda\int F d\mu + n\lambda^2/2}$$

for every $|\lambda| \le 1$. By Chebyshev's inequality,

$$\mu\big(\{F \ge \textstyle\int F d\mu + r\}\big) \le \mathrm{e}^{-r^2/2n} \tag{5.27}$$

however only for $0 \leq r \leq n$. But since $\|F\|_{\mathrm{Lip}} \leq 1$ for the Hamming metric, $|F - \int F d\mu| \leq n$ so that (5.27) actually holds true for every $r \geq 0$.

It might be further observed that if F is convex on $[0,1]^n$, then

$$\left|D_i F(x)\right|^2 \leq \left|\partial_i F(x)\right|^2 + \left|\partial_i F\big(s_i(x)\big)\right|^2$$

where ∂_i is the partial derivative of F along the i-th coordinate. Then

$$\mathrm{Ent}_\mu(\mathrm{e}^{\lambda F}) \leq \int \left|\nabla(\mathrm{e}^{\lambda F/2})\right|^2 d\mu \leq \frac{\lambda^2}{4}\int |\nabla F|^2 \mathrm{e}^{\lambda F} d\mu, \qquad \lambda \geq 0.$$

Together with the Herbst argument, we thus recover in this way Theorem 5.9 with improved constants but only for uniform measure on $\{0,1\}^n$.

Now, for $0 \leq p \leq 1$, let μ_p^n be the product measure on $\{0,1\}^n$ when each factor is endowed with the Bernoulli measure $p\delta_1 + q\delta_0$, $q = 1-p$, with probability of success p. Then the constant in the logarithmic Sobolev inequality (5.26) for μ_p^n does not behave nicely as a function of p. Indeed (see [SC], [An]),

$$\mathrm{Ent}_{\mu_p^n}(f^2) \leq C_p \sum_{i=1}^n \int |D_i f|^2 d\mu_p^n \tag{5.28}$$

where

$$C_p = pq\,\frac{\log p - \log q}{p-q}\,.$$

In particular, $C_p/pq \to +\infty$ as $p \to 0$ or 1. This prevents the use of (5.28) in any limit theorem as $p \to 0$, such as Poissonian limits. There is of course a good reason for that, namely that Poisson measures do not satisfy standard logarithmic Sobolev inequalities with respect to the gradient D of (5.25). Indeed, denote by π_θ the Poisson measure on $\mathbb{N}$ with parameter $\theta > 0$ and assume that, for some constant $C > 0$, and all f, say bounded, on $\mathbb{N}$,

$$\mathrm{Ent}_{\pi_\theta}(f^2) \leq C\int |Df|^2 d\pi_\theta \tag{5.29}$$

where we recall that $Df(x) = f(x+1) - f(x)$, $x \in \mathbb{N}$. Apply (5.29) to the indicator function of the interval $[k+1,\infty)$ for each $k \in \mathbb{N}$. We get

$$-\pi_\theta\big([k+1,\infty)\big)\log \pi_\theta\big([k+1,\infty)\big) \leq C\pi_\theta\big(\{k\}\big)$$

which is clearly impossible as k goes to infinity.

There is however an alternative version of the logarithmic Sobolev inequality on the two-point space for which Poissonian limits may be performed. Recall the product Bernoulli measure μ_p^n with parameter p.

Lemma 5.18. *For any non-negative function f on $\{0,1\}^n$,*

$$\mathrm{Ent}_{\mu_p^n}(f) \leq pq\int \sum_{i=1}^n D_i f D_i(\log f)\, d\mu_p^n.$$

When the gradients D_i satisfy the chain rule formula, the form of the logarithmic Sobolev inequality of Lemma 5.18 is equivalent by integration by parts to the standard form. For example, the logarithmic Sobolev inequality for the canonical Gaussian measure γ on $\mathbb{R}^n$ is equivalent to saying that for any smooth non-negative function f on $\mathbb{R}^n$,

$$\mathrm{Ent}_\gamma(f) \leq \frac{1}{2} \int \nabla f \cdot \nabla(\log f) d\gamma.$$

The proof of Lemma 5.18 reduces as usual to dimension one, in which case it amounts to establishing the inequality

$$\begin{aligned} pu \log u + qv \log v - (pu + qv) \log(pu + qv) \\ \leq pq(u - v)(\log u - \log v) \end{aligned}$$

for any $u, v \geq 0$ (see [Wu], [An]).

Due to the constant pq in the logarithmic Sobolev inequality of Lemma 5.18, we may perform a Poissonian limit as $p \to 0$. Take ϕ on $\mathbb{N}$ such that

$$0 < \varepsilon \leq \phi \leq 1/\varepsilon < \infty$$

and apply Lemma 5.18 to

$$f(x) = f(x_1, \ldots, x_n) = \phi(x_1 + \cdots + x_n),$$

$x = (x_1, \ldots, x_n) \in \{0, 1\}^n$, with $p = \frac{\tau}{n}$, $\tau > 0$ (for every n large enough). Then, setting $S_n = x_1 + \cdots + x_n$,

$$\begin{aligned} &\sum_{i=1}^{n} D_i f D_i(\log f) \\ &\quad = (n - S_n)\big[\phi(S_n + 1) - \phi(S_n)\big]\big[\log \phi(S_n + 1) - \log \phi(S_n)\big] \\ &\qquad + S_n\big[\phi(S_n) - \phi(S_n - 1)\big]\big[\log \phi(S_n) - \log \phi(S_n - 1)\big]. \end{aligned}$$

The distribution of S_n under $\mu^n_{\tau/n}$ converges to π_τ. Using the bounds

$$0 < \varepsilon \leq \phi \leq 1/\varepsilon < \infty,$$

it follows from the preceding that

$$\frac{\tau}{n}\Big(1 - \frac{\tau}{n}\Big) \int \sum_{i=1}^{n} D_i f D_i(\log f) d\mu^n_{\tau/n} \to \int D\phi D(\log \phi) d\pi_\tau$$

as $n \to \infty$. We may therefore state the following corollary.

Corollary 5.19. *For any non-negative function f on $\mathbb{N}$,*

$$\mathrm{Ent}_{\pi_\tau}(f) \leq \tau \int Df D(\log f) d\pi_\tau$$

where we recall that here $Df(x) = f(x + 1) - f(x)$, $x \in \mathbb{N}$.

The example of $f(x) = \mathrm{e}^{-\kappa x}$, $x \in \mathbb{N}$, as $\kappa \to \infty$ shows that one cannot expect a better factor of τ in the preceding corollary.

To conclude this section, we deduce concentration inequalities for measures μ on $\mathbb{N}$, for simplicity, satisfying a logarithmic Sobolev inequality, such as the Poisson measure π_τ.

Corollary 5.20. *Let μ be a probability measure on $\mathbb{N}$ such that for some constant $C > 0$ and all non-negative (bounded) functions f on $\mathbb{N}$,*

$$\mathrm{Ent}_\mu(f) \le C \int DfD(\log f)d\mu.$$

Then, if F is such that $|F(x+1) - F(x)| \le 1$ for all $x \in \mathbb{N}$, we have $\int |F| d\mu < \infty$ and, for every $r \ge 0$,

$$\mu(\{F \ge \textstyle\int F d\mu + r\}) \le \exp\Big(-\frac{r}{4}\log\Big(1 + \frac{r}{2C}\Big)\Big).$$

The numerical constants in Corollary 5.20 have no reason to be sharp. However, the order of growth is optimal as shown by the example of Poisson measure itself. More precisely, the tail of the Lipschitz function F is Gaussian for the small values of r and Poissonian for the large values (with respect to C).

Proof. It follows the common pattern of proofs in this chapter. We apply the logarithmic Sobolev inequality of Corollary 5.20 to $f = \mathrm{e}^{\lambda F}$ with F say bounded to start with. The gradient D is not local; however, since $|F(x+1) - F(x)| \le 1$ for all $x \in \mathbb{N}$, for every $\lambda \ge 0$,

$$\big[\lambda F(x+1) - \lambda F(x)\big]\big[\mathrm{e}^{\lambda F(x+1)} - \mathrm{e}^{\lambda F(x)}\big] \le \lambda\, \mathrm{e}^{\lambda} \mathrm{e}^{\lambda F(x)}.$$

Hence, with the usual notation,

$$\mathrm{Ent}_\mu(\mathrm{e}^{\lambda F}) \le \lambda \mathrm{e}^{\lambda} \Lambda(\lambda).$$

Integration of this differential inequality together with Chebyschev's inequality easily yields the result. Corollary 5.20 is established. □

According to this proof, the preceding logarithmic Sobolev inequalities are part of the family of modified logarithmic Sobolev inequalities investigated in Section 5.3, with a function $\beta(\rho)$ of the order of $\mathrm{e}^{2\rho}$, $\rho \ge 0$. Following Proposition 5.13 it may thus be tensorized in terms of two distinct norms on the gradients. The following statement is then an easy consequence of this observation. It extends to Lipschitz functions known inequalities for sums of independent Bernoulli or Poisson random variables. We let $(e_1, \ldots, e_n)$ be the canonical basis of $\mathbb{R}^n$.

Corollary 5.21. *Let μ be some measure on $\mathbb{N}$. Assume that for every f on $\mathbb{N}$ with $\sup_{x \in \mathbb{N}} |Df(x)| \le \rho$,*

$$\mathrm{Ent}_\mu(\mathrm{e}^f) \le \beta(\rho) \int |Df|^2 \mathrm{e}^f d\mu$$

where, as a function of $\rho \geq 0$, $\beta(\rho) \leq \kappa\, \mathrm{e}^{\delta\rho}$ *for some* $\kappa, \delta > 0$. *Denote by* μ^n *the* n*-fold product measure of* μ *on* $\mathbb{N}^n$. *Let* F *be a function on* $\mathbb{N}^n$ *such that, for every* $x \in \mathbb{N}^n$,

$$\sum_{i=1}^{n} \big|F(x+e_i) - F(x)\big|^2 \leq a^2 \quad \text{and} \quad \max_{1 \leq i \leq n} \big|F(x+e_i) - F(x)\big| \leq b.$$

Then $\int |F| d\mu^n < \infty$ *and, for every* $r \geq 0$,

$$\mu^n\big(\{F \geq \textstyle\int F d\mu^n + r\}\big) \leq \exp\bigg(-\frac{r}{2\delta b} \log\Big(1 + \frac{bdr}{4\kappa a^2}\Big)\bigg).$$

5.5 Covariance identities

In this last section, we consider yet another approach to some of the concentration properties described in this chapter and the preceding ones based on covariance identities. We only present the basic principle in the Gaussian case, referreing to [B-G-H], [Hou], [Pa] for extensions to Poisson and infinitely divisible measures.

Recall the Ornstein-Uhlenbeck semigroup (5.3),

$$P_t f(x) = \int f\big(\mathrm{e}^{-t}x + (1-\mathrm{e}^{-2t})^{1/2}y\big) d\gamma(y), \quad t \geq 0,\ x \in \mathbb{R}^n,$$

with symmetric and invariant measure the canonical Gaussian measure γ on $\mathbb{R}^n$. Let f and g be smooth functions on $\mathbb{R}^n$. Using that $\frac{d}{dt} P_t g = \mathrm{L} P_t g$ where L is the second order differential operator $\Delta - x \cdot \nabla$, we may write as in Sections 2.3 and 5.1 that

$$\begin{aligned}
\mathrm{Cov}_\gamma(f,g) &= \int f\big(g - \textstyle\int g d\gamma\big) d\gamma d\gamma \\
&= -\int_0^\infty \int f\, \mathrm{L} P_t g \, d\gamma dt \\
&= \int_0^\infty \int \nabla f \cdot \nabla P_t g \, d\gamma dt.
\end{aligned}$$

Coming back to the expression of $P_t g$,

$$\mathrm{Cov}_\gamma(f,g) = \int_0^\infty \iint \nabla f(x) \cdot \nabla g\big(\mathrm{e}^{-t}x + (1-\mathrm{e}^{-2t})^{1/2}y\big) d\gamma(x) d\gamma(y)\, dt.$$

Therefore, if ϖ_t is the Gaussian measure on $\mathbb{R}^n \times \mathbb{R}^n$ which is the image measure of $\gamma \otimes \gamma$ by the map

$$(x,y) \mapsto \big(x, \mathrm{e}^{-t}x + (1-\mathrm{e}^{-2t})^{1/2}y\big)$$

and if ϖ denotes the probability measure $\int_0^\infty \varpi_t dt$, we get that

$$\mathrm{Cov}_\gamma(f,g) = \iint \nabla f(x) \cdot \nabla g(y) d\varpi(x,y). \tag{5.30}$$

It is of course assumed in the preceding argument that f and g have all the regularity and integrability properties needed to justify differentiation and the use of Fubini's theorem.

The representation formula (5.30) (which may be shown to hold in more general situations) is a simple tool to reach measure concentration. The following is an immediate consequence of (5.30).

Proposition 5.22. *For every smooth enough functions f and g on $\mathbb{R}^n$ such that f is 1-Lispchitz,*

$$\mathrm{Cov}_\gamma(f, g) \le \int |\nabla g| d\gamma.$$

Now, Proposition 5.22 may be used as logarithmic Sobolev inequalities to produce measure concentration with a modification of the Herbst argument. Indeed, given F 1-Lipschitz and with mean zero with respect to γ, apply Proposition 5.22 to $f = F$ and $g = \mathrm{e}^{\lambda F}$, $\lambda \ge 0$. Then

$$\int F \mathrm{e}^{\lambda F} d\gamma = \mathrm{Cov}_\gamma(F, \mathrm{e}^{\lambda F}) \le \lambda \int |\nabla F| \, \mathrm{e}^{\lambda F} d\gamma \le \lambda \int \mathrm{e}^{\lambda F} d\gamma.$$

For the function $J(\lambda) = \log \int \mathrm{e}^{\lambda F} d\gamma$, $\lambda \ge 0$, we thus have the differential inequality $J'(\lambda) \le \lambda$. Since $J(0) = 0$, we conclude that $J(\lambda) \le \frac{\lambda^2}{2}$. Hence

$$\int \mathrm{e}^{\lambda F} d\gamma \le \mathrm{e}^{\lambda^2/2}.$$

We thus recover from the covariance identity (5.30) the concentration properties of Gaussian measures.

This type of argument may be, on one hand, used to sharpen some of the deviation inequalities for Gaussian measures, and, on the other hand, be extended to various kinds of distributions including binomial, Poisson and infinitely divisible laws. We refer to [B-G-H], [H-PA-S], [Hou] and [Pa].

Notes and Remarks

Logarithmic Sobolev inequalities were introduced by L. Gross [Gros1] as the infinitesimal version of hypercontractivity in quantum field theory (see [Gros2] and references therein). They soon became a tool of fundamental importance in infinite dimensional analysis, with increasing activity [Gros2], both on the side of lattice spin systems and Gibbs measures in statistical mechanics ([Str1], [G-Zeg], [Roy], etc.), and of path and loop spaces in infinite dimensional stochastic analysis [Hs]. The smoothing property of hypercontractivity of the associated heat semigroup is there a tool of powerful interest in convergence to equilibrium and uniqueness theorems. Applications of logarithmic Sobolev inequalities to convergence to equilibrium of a finite state Markov chain are presented in [D-SC], [SC]. A pedestrian introduction to logarithmic Sobolev inequalities is the reference [An].

One of the early questions on logarithmic Sobolev inequalities was to determine conditions on measures μ on $\mathbb{R}^n$ to satisfy a logarithmic Sobolev inequality

$$\mathrm{Ent}_\mu(f^2) \le 2C \int |\nabla f|^2 d\mu$$

for all smooth enough functions f. To this question raised by L. Gross, I. Herbst (in an unpublished letter to L. Gross) found as a necessary condition the fact that

$$\int e^{\rho|x|^2} d\mu(x) < \infty$$

for some $\rho > 0$ small enough. Herbst's argument is presented in [Da-S], however, with a small error in the argument. This was settled in the paper [A-M-S], which revived the interest in the Gross question and gave rise to several subsequent contributions [A-S], [Le2], [G-R], [Rot2], etc. In particular, Herbst's argument as a Laplace bound of concentration spirit is exposed in [Le2] and is presented this way as Theorem 5.3.

The Herbst argument motivated the investigation [Le4] of concentration inequalities for product measures by the entropic method. The results of Section 5.2 are taken from this reference. Some sharper numerical constants may be found in [Le4], [Bob1], [Mas1], etc.

Modified logarithmic Sobolev inequalities and their application to the concentration properties of products of the exponential distribution were investigated in [Bo-L1] from which the results of Section 5.3 are taken. The corresponding results for products of Markov chains are discussed in [H-T]. Concentration between Poincaré and logarithmic Sobolev inequalities with application to concentration of log-concave measures is investigated in [L-O] (see also [Bar1]) through the family of functional inequalities (5.24).

The discrete setting is studied in a number of references including [A-S], [G-R], etc. The results of Section 5.4 are taken from [Wu] and [Bo-L2] (a different energy is however used in the latter) where Poissonian bounds are analyzed (see also the references on Section 5.5).

Most of the results of this chapter are taken from the survey [Le5] where the reader will find further details and references. It must also be emphasized that a great deal of the intense activity on logarithmic Sobolev inequalities in the recent years has been dealing with logarithmic Sobolev inequalities for infinite particle systems (lattice spin systems and Gibbs states in statistical mechanics), especially with the work of D. Stroock and B. Zegarlinski (cf. [Str1], [G-Zeg], [Roy], etc.). For recent developments for unbounded spin systems, see [Yos], [B-H], [Le7]. These are examples of dimension free logarithmic Sobolev inequalities which are highly non-product measures. Together with the Herbst argument, these results thus yield new concentration properties far away from product spaces. Another direction to non-product measures will be alluded to in Section 6.3 of the next chapter. For applications to concentration on path spaces (extending Wiener spaces), see [Hs], [Le5], [A-L], [Ho-P], etc.

The result and method of Section 5.5 are taken from the works [B-G-H], [H-PA-S], [Hou], [Pa] and [Ho-P], to which we refer for further developments for infinitely divisible distributions. It is shown in [B-G-H] that the method can be used to yield sharper bounds and constants.

6. TRANSPORTATION COST INEQUALITIES

In this chapter, we investigate a third description of concentration. After the geometric description of the concentration function itself, and the Laplace bounds related to logarithmic Sobolev inequalities, we turn to a dual formulation of the latter in terms of distances between measures. This approach was put forward by K. Marton. While equally suited to product measures, it seems a convenient tool to investigate concentration in some dependent structures such as contractive Markov chains. After describing information inequalities and their relation to concentration, we study quadratic transportation cost inequalities. While these are actually consequences of logarithmic Sobolev inequalities, they have an interest in their own. In the last section, we present proofs of some of the geometric inequalities of Sections 4.2 and 4.3 with the transportation cost approach, and discuss some extensions to non-product measures.

6.1 Information inequalities and concentration

To introduce to the topic of this chapter, let us start with the classical Pinsker-Csizsar-Kullback inequality (cf. [Pin]) that indicates that whenever μ and ν are two probability measures on the Borel sets of a metric space (X, d), then

$$\|\mu - \nu\|_{\mathrm{TV}} \leq \sqrt{\frac{1}{2}\,\mathrm{H}(\nu\,|\,\mu)}\,. \tag{6.1}$$

Here $\|\cdot\|_{\mathrm{TV}}$ denotes the total variation distance, whereas $\mathrm{H}(\nu\,|\,\mu)$ is the *relative entropy* of ν with respect to μ defined by

$$\mathrm{H}(\nu\,|\,\mu) = \mathrm{Ent}_\mu\Big(\frac{d\nu}{d\mu}\Big) = \int \log\frac{d\nu}{d\mu}\,d\nu$$

whenever ν is absolutely continuous with respect to μ with Radon-Nikodym derivative $\frac{d\nu}{d\mu}$, and $+\infty$ if not. Inequalities such as (6.1) have often been considered in *information theory*.

That such an inequality is related to concentration properties may be shown in the following way. Given a metric space (X, d) and two Borel probability measures μ and ν on X, consider the Wasserstein distance between μ and ν,

$$\mathrm{W}_1(\mu,\nu) = \inf \iint d(x,y)\,d\pi(x,y)$$

where the infimum runs over all probability measures π on the product space $X \times X$ with marginals μ and ν having a finite first moment. The total variation distance corresponds to the trivial metric. Given μ, consider then the inequality

$$\mathrm{W}_1(\mu,\nu) \leq \sqrt{2C\mathrm{H}(\nu\,|\,\mu)} \tag{6.2}$$

for some $C > 0$ and every ν. Let A and B be Borel sets with $\mu(A), \mu(B) > 0$, and consider the conditional probabilities $\mu_A = \mu(\cdot\,|A)$ and $\mu_B = \mu(\cdot\,|B)$. By the triangle inequality for W_1 and (6.2),

$$\begin{aligned}\mathrm{W}_1(\mu_A,\mu_B) &\leq \mathrm{W}_1(\mu,\mu_A) + \mathrm{W}_1(\mu,\mu_B)\\ &\leq \sqrt{2C\,\mathrm{H}(\mu_A\,|\,\mu)} + \sqrt{2C\,\mathrm{H}(\mu_B\,|\,\mu)}\\ &= \sqrt{2C\log\frac{1}{\mu(A)}} + \sqrt{2C\log\frac{1}{\mu(B)}}\,.\end{aligned} \tag{6.3}$$

Now, all measures with marginals μ_A and μ_B must be supported on $A \times B$, so that, by definition of W_1,

$$\mathrm{W}_1(\mu_A,\mu_B) \geq d(A,B) = \inf\big\{d(x,y); x\in A, y\in B\big\}.$$

Then (6.3) implies a concentration inequality. Given A and B in X such that $d(A,B) \geq r > 0$, we get

$$r \leq \sqrt{2C\log\frac{1}{\mu(A)}} + \sqrt{2C\log\frac{1}{1-\mu(A_r)}} \tag{6.4}$$

where we recall that $A_r = \{x\in X; d(x,A) < r\}$. Inequality (6.4) appears as a symmetric form of concentration. If, say, $\mu(A) \geq \frac{1}{2}$,

$$r \leq \sqrt{2C\log 2} + \sqrt{2C\log\frac{1}{1-\mu(A_r)}}$$

so that, whenever $r \geq 2\sqrt{2C\log 2}$ for example,

$$1-\mu(A_r) \leq \mathrm{e}^{-r^2/8C}.$$

The same applies for Wasserstein functionals with respect to some cost function $\tilde{c} : X\times X \to \mathbb{R}_+$ (cf. [R-R]) defined by

$$\mathrm{W}_{\tilde{c}}(\mu,\nu) = \inf \iint \tilde{c}(x,y)d\pi(x,y)$$

where the infimum is running over all probability measures π on the product space $X\times X$ with marginals μ and ν such that $\tilde{c}$ is integrable with respect to π. Given μ, we may consider the *transportation cost inequality*

$$\mathrm{W}_{\tilde{c}}(\mu,\nu) \leq \mathrm{H}(\nu\,|\,\mu) \tag{6.5}$$

for every ν. Arguing as above shows that for every Borel sets A and B in X,

$$\inf_{x\in A, y\in B} \tilde{c}(x,y) \leq \sqrt{2\log\frac{1}{\mu(A)}} + \sqrt{2\log\frac{1}{1-\mu(A_r)}}\,. \tag{6.6}$$

We thus reach in this way concentration properties of the family of sets of Proposition 1.18.

It actually turns out that the transportation cost inequality (6.2) is equivalent to the normal Laplace bound deduced in the preceding chapter from logarithmic Sobolev inequalities, and may be thought of as its dual version. We indeed have the following result. Recall the Laplace functional (Section 1.6) of (X,d,μ),

$$\mathrm{E}_{(X,d,\mu)}(\lambda) = \sup \int \mathrm{e}^{\lambda F} d\mu, \quad \lambda \geq 0,$$

where the supremum runs over all 1-Lipschitz mean zero functions $F : X \to \mathbb{R}$.

Proposition 6.1. *Let μ be a Borel probability measure on a metric space (X,d). Then*

$$\mathrm{W}_1(\mu,\nu) \leq \sqrt{2C\mathrm{H}(\nu\,|\,\mu)} \tag{6.7}$$

for some $C > 0$ and all ν if and only if

$$\mathrm{E}_{(X,d,\mu)}(\lambda) \leq \mathrm{e}^{C\lambda^2/2}, \quad \lambda \geq 0. \tag{6.8}$$

Proof. By the Monge-Kantorovitch-Rubinstein dual characterization (cf. [Dud], [Ra2]) of the Wasserstein distance,

$$\mathrm{W}_1(\mu,\nu) = \sup \left[\int g d\nu - \int f d\mu\right]$$

where the supremum is running over all bounded measurable functions f and g such that

$$g(x) \leq f(y) + d(x,y)$$

for every $x, y \in X$. Under (6.7),

$$\int g d\nu - \int f d\mu \leq \sqrt{2C\,\mathrm{Ent}_\mu\Big(\frac{d\nu}{d\mu}\Big)}\,,$$

or, equivalently, for every $\lambda > 0$,

$$\int g d\nu - \int f d\mu \leq \frac{C\lambda}{2} + \frac{1}{\lambda}\,\mathrm{Ent}_\mu\Big(\frac{d\nu}{d\mu}\Big).$$

Set $\phi = \frac{d\nu}{d\mu}$. The preceding indicates that

$$\int \psi\phi d\mu \leq \mathrm{Ent}_\mu(\phi)$$

where $\psi = \lambda g - \lambda \int f d\mu - C\lambda^2/2$. Since this inequality holds for every choice of ϕ (i.e. ν), applying it to $\phi = \mathrm{e}^{\psi} / \int \mathrm{e}^{\psi} d\mu$ yields that $\log \int \mathrm{e}^{\psi} d\mu \leq 0$. In other words,

$$\int \mathrm{e}^{\lambda g} d\mu \leq \mathrm{e}^{\lambda \int f d\mu + C\lambda^2/2}.$$

When F is Lipschitz with $\|F\|_{\mathrm{Lip}} \leq 1$, one may choose $F = g = f$ so that the latter amounts to (6.8). Since

$$\mathrm{Ent}_{\mu}(\phi) = \sup \int \phi \psi d\mu$$

where the supremum is running over all ψ's such that $\int \mathrm{e}^{\psi} d\mu \leq 1$, the preceding argument clearly indicates that (6.8) is actually equivalent to (6.7). The proof of Proposition 6.1 is complete. □

The same may be proved on the alternative (more classical and easily equivalent) characterization of W_1 as

$$\mathrm{W}_1(\mu, \nu) = \sup \left[\int F d\mu - \int F d\nu \right]$$

where the supremum is running over all 1-Lipschitz functions F on (X, d).

The general form of the dual Monge-Kantorovitch-Rubinstein representation theorem indicates that (cf. [Ra1], [Ra2], [R-R]) for Borel probability measures μ and ν on a metric space (X, d),

$$\mathrm{W}_{\tilde{c}}(\mu, \nu) = \inf \iint \tilde{c}(x, y) d\pi(x, y) = \sup \left[\int g d\nu - \int f d\mu \right] \tag{6.9}$$

where the supremum is over all pairs (g, f) of bounded measurable functions (or respectively ν and μ-integrable) such that for all x, y in X,

$$g(x) \leq f(y) + \tilde{c}(x, y).$$

Here $\tilde{c}$ is upper semicontinuous, π-integrable and such that

$$\tilde{c}(x, y) \leq a(x) + b(y)$$

for some measurable functions a and b. On $\mathbb{R}^n$, the supremum on the right-hand side of (6.9) may be taken over smaller classes of smooth functions, such as bounded Lipschitz or so on. On the basis of this description, it is a mere exercise to repeat the proof of Proposition 6.1 to come to the following. We recall from (1.27) the infimum-convolution $Q_{\tilde{c}} f$ with cost $\tilde{c}$ of a given function f,

$$Q_{\tilde{c}} f(x) = \inf_{y \in X} \big[f(y) + \tilde{c}(x, y) \big], \quad x \in X.$$

Proposition 6.2. *Let $\tilde{c}$ be an admissible (for (6.9) to hold) cost function on $X \times X$ and let μ be a Borel probability measure on X. Then*

$$\mathrm{W}_{\tilde{c}}(\mu, \nu) \leq \mathrm{H}\big(\nu \,|\, \mu\big) \tag{6.10}$$

for all ν if and only if for all bounded measurable functions f on $\mathbb{R}^n$,

$$\int \mathrm{e}^{Q_{\tilde{c}}f} d\mu \leq \mathrm{e}^{\int f d\mu}. \tag{6.11}$$

The infimum-convolution inequality (6.11) of Proposition 6.2 has of course to be compared with the one introduced in Section 1.6 that reads

$$\int \mathrm{e}^{Q_{\tilde{c}}f} d\mu \int \mathrm{e}^{-f} d\mu \leq 1. \tag{6.12}$$

By Jensen's inequality $\int \mathrm{e}^{-f} d\mu \geq \mathrm{e}^{\int f d\mu}$, so that (6.12) is stronger than (6.11). As we have seen in Proposition 1.18 and (6.6) the infimum convolution inequalities (6.11) and (6.12) both produce measure concentration for μ with respect to the cost $\tilde{c}$. Proposition 6.2 is actually a bridge between the methods developed in Section 1.6 (and Sections 4.4 and 4.5) and the transportation cost inequalities. As announced, it may be used for example to yield a very simple proof of the product property of transportation cost inequalities along the lines of Proposition 1.19.

Proposition 6.3. *Let $P = \mu_1 \otimes \cdots \otimes \mu_n$ be a product probability measure on the Borel sets of a product space $X = X_1 \times \cdots \times X_n$ of metric spaces (X_i, d_i), $i = 1, \ldots, n$. Assume that each μ_i, $i = 1, \ldots, n$, satisfies a quadratic transportation cost inequality*

$$\mathrm{W}_{\tilde{c}_i}(\mu_i, \nu) \leq \mathrm{H}(\nu \,|\, \mu_i)$$

for every ν_i on X_i for some cost function $\tilde{c}_i$. Then,

$$\mathrm{W}_{\tilde{c}}(P, R)) \leq \mathrm{H}(R \,|\, P)$$

for every probability measure R on X with respect to the cost $\tilde{c} = \sum_{i=1}^n \tilde{c}_i$.

Proof. By Proposition 6.2, we may follow the argument put forward in Proposition 1.19. By induction, it is enough to consider the case $n = 2$. Let f be bounded on $X_1 \times X_2$, and for $x_2 \in X_2$, set $g(x_2) = \log \int \mathrm{e}^{Q_{\tilde{c}_1} f^{x_2}} d\mu_1$ where $f^{x_2}(\cdot) = f(\cdot, x_2)$. As we already saw in Proposition 1.19,

$$\int \mathrm{e}^{Q_{\tilde{c}}f} d\mu_1 d\mu_2 \leq \int \mathrm{e}^{Q_{\tilde{c}_2}g} d\mu_2.$$

By (6.11) applied to g,

$$\int \mathrm{e}^{Q_{\tilde{c}_2}g} d\mu_2 \leq \mathrm{e}^{\int g d\mu_2}$$

and by (6.11) applied to f^{x_2} for every x_2,

$$g(x_2) = \log \int \mathrm{e}^{Q_{\tilde{c}_1} f^{x_2}} d\mu_1 \leq \int f^{x_2} d\mu_1.$$

The claim follows. □

6.2 Quadratic transportation cost inequalities

As we have seen in Proposition 6.3, transportation cost inequalities do share tensorization properties as logarithmic Sobolev inequalities and may be used, for some appropriate cost functions, to describe dimension free concentration properties.

Consider indeed metric spaces (X_i, d_i), $i = 1, \dots, n$, equipped with probability measures μ_i, $i = 1, \dots, n$. Let P denote the product measure $P = \mu_1 \otimes \cdots \otimes \mu_n$ on the product space $X = X_1 \times \cdots \times X_n$. By the product property of Laplace functionals (Proposition 1.15) and Proposition 6.1, or Proposition 6.3, if each measure μ_i satisfies a transportation cost inequality

$$\mathrm{W}_1(\mu_i, \nu_i) \leq \sqrt{2C\mathrm{H}(\nu_i \,|\, \mu_i)}$$

for all ν_i on X_i, $i = 1, \dots, n$, then the product measure P satisfies the transportation cost inequality

$$\mathrm{W}_1(P, R) \leq \sqrt{2Cn\mathrm{H}(R \,|\, P)}$$

for every probability measure R on the Cartesian product space equipped with the ℓ^1-metric. As discussed earlier, the drawback of the ℓ^1-metric is that it highly takes into account the number of coordinates in the product space. However, it allows us to recover once more concentration with respect to the Hamming metric. Indeed, starting from the Pinsker-Csizsar-Kullback inequality (6.1), for any product probability measure P on X equipped with the Hamming metric,

$$\mathrm{W}_1(P, R) \leq \sqrt{\frac{n}{2}\,\mathrm{H}(R \,|\, P)}$$

which thus produces concentration by Proposition 6.1.

Following the logarithmic Sobolev approach of Chapter 5, it is however more fruitful, in order to reach dimension free concentration properties, to think in terms of a quadratic cost. To this task, let us restrict ourselves to the case of $\mathbb{R}^n$ with the Euclidean norm $|\cdot|$. Given a probability measure μ on the Borel sets of $\mathbb{R}^n$, say that it satisfies a *quadratic transportation cost inequality* whenever there exists a constant $C > 0$ such that for all probability measures ν,

$$\mathrm{W}_2(\mu, \nu) \leq \sqrt{C\,\mathrm{H}(\nu \,|\, \mu)}. \tag{6.13}$$

Here W_2 is the Wasserstein distance with quadratic cost

$$\mathrm{W}_2(\mu, \nu)^2 = \inf \iint \frac{1}{2}\,|x - y|^2 d\pi(x, y)$$

where the infimum is running over all probability measures π on $\mathbb{R}^n \times \mathbb{R}^n$ with respective marginals μ and ν. (The infimum in W_2 is finite as soon as μ and ν have finite second moment which we shall always assume.)

It is clear by Jensen's inequality that the quadratic transportation cost inequality is stronger than the W_1 transportation cost inequality considered in Section 6.1.

Since the cost in W_2 is given by $\tilde{c}(x, y) = \tilde{c}(x - y)$ with

$$\tilde{c}(x) = \frac{1}{2}\,|x|^2 = \frac{1}{2}\sum_{i=1}^{n} x_i^2, \quad x = (x_1, \dots, x_n) \in \mathbb{R}^n,$$

Proposition 6.3 applies to show that a quadratic transportation cost inequality tensorizes. In particular, if each μ_i satisfies a quadratic transportation cost inequality with constants $C_i > 0$, $i = 1, \ldots, n$, then the product measure $P = \mu_1 \otimes \cdots \otimes \mu_n$ will satisfy a quadratic transportation cost inequality with constant $C = \max_{1 \leq i \leq n} C_i$ (compare with Corollary 5.7). Moreover, Proposition 6.2 for the quadratic cost reads as follows.

Corollary 6.4. *Let μ be a Borel probability measure on $\mathbb{R}^n$. Then*

$$\mathrm{W}_2(\mu, \nu) \leq \sqrt{C\mathrm{H}(\nu \,|\, \mu)} \tag{6.14}$$

for some $C > 0$ and all ν if and only if for all bounded measurable functions f on $\mathbb{R}^n$,

$$\int \mathrm{e}^{Q_{1/C} f} d\mu \leq \mathrm{e}^{\int f d\mu} \tag{6.15}$$

where $Q_c f$, $c > 0$, is the infimum-convolution of f with the quadratic cost $\tilde{c}(x, y) = \frac{c}{2} |x - y|^2$, that is,

$$Q_c f(x) = \inf_{y \in \mathbb{R}^n} \Big[f(y) + \frac{c}{2} |x - y|^2 \Big], \quad x \in \mathbb{R}^n.$$

As in Section 1.6, observe, for the matter of comparison with Proposition 6.1, that whenever F is Lipschitz,

$$Q_c F \geq F - \frac{1}{2c} \|F\|_{\mathrm{Lip}}^2.$$

The quadratic transportation cost inequalities thus turn out to be a useful tool in the investigation of dimension free concentration properties similar to what has been developed for logarithmic Sobolev inequalities in the preceding chapter. In a sense, the quadratic transportation cost inequalities may be thought of as dual to logarithmic Sobolev inequalities.

Corollary 6.4 furthermore emphasizes the usefulness of Brunn-Minkowski inequalities in this context, which may be compared to what was developed in Section 2.2. Let μ be a Borel probability measure on $\mathbb{R}^n$ with density e^{-U} with respect to Lebesgue measure. Assume that U is strictly convex in the sense that, for some $c > 0$ and every $\theta \in [0, 1]$, $x, y \in \mathbb{R}^n$,

$$\theta U(x) + (1 - \theta) U(y) - U\big(\theta x + (1 - \theta) y\big) \geq \frac{c}{2} \theta(1 - \theta) |x - y|^2. \tag{6.16}$$

If the potential U is twice continuously differentiable, this amounts to $\mathrm{Hess}\,(U)(x) \geq c\,\mathrm{Id}$ (as symmetric matrices) uniformly in $x \in \mathbb{R}^n$.

Theorem 6.5. *Let $d\mu = \mathrm{e}^{-U} dx$ where U satisfies (6.16) with constant $c > 0$. Then, for every probability measure ν on $\mathbb{R}^n$,*

$$\mathrm{W}_2(\mu, \nu) \leq \sqrt{\frac{1}{c} \mathrm{H}(\nu \,|\, \mu)}\,.$$

Before turning to the proof of Theorem 6.5, note that it contains the case of the canonical Gaussian measure on $\mathbb{R}^n$ with $c = 1$. This case may be proved alternatively starting from dimension one and using induction over the coordinates (see [Tal10]. As shown in [Bl1], more generally, an alternative proof of Theorem 6.5 may be given using the Brenier-McCann theorem [Bre], [MC] about monotone measure preserving maps.

Note furthermore that by Corollary 6.4, under the hypotheses of Theorem 6.5,

$$\int e^{Q_c f} d\mu \leq e^{\int f d\mu}$$

for every bounded measurable function f while Theorem 2.15 of Chapter 2 shows that

$$\int e^{Q_{c/2} f} d\mu \int e^{-f} d\mu \leq 1.$$

(However Theorem 2.15 only uses (6.16) with $\theta = \frac{1}{2}$.) We thus reach with the quadratic cost transportation inequalities somewhat sharper bounds under somewhat sharper hypotheses.

Proof. Given $\theta \in [0, 1]$ and $x, y \in \mathbb{R}^n$, let

$$L_\theta(x, y) = \frac{1}{\theta(1-\theta)} \Big[\theta U(x) + (1-\theta)U(y) - U\big(\theta x + (1-\theta)y\big)\Big].$$

We apply the functional form of the Brunn-Minkowski theorem (Section 2.2). Set

$$u(x) = e^{-(1-\theta)f(x)-U(x)}, \quad v(y) = e^{\theta g(y)-U(y)}, \quad w(z) = e^{-U(z)}.$$

We get

$$1 \geq \left(\int e^{-(1-\theta)f} d\mu\right)^{\theta} \left(\int e^{\theta g} d\mu\right)^{1-\theta} \tag{6.17}$$

provided the functions f and g satisfy

$$g(y) \leq f(x) + L_\theta(x, y), \quad x, y \in \mathbb{R}^n. \tag{6.18}$$

Given f, the optimal function $g = L_\theta f$ in (6.18) is defined by

$$L_\theta f(y) = \inf_{x \in \mathbb{R}^n} \big[g(x) + L_\theta(x, y)\big]$$

so that (6.17) becomes

$$\left(\int e^{-(1-\theta)f} d\mu\right)^{1/(1-\theta)} \left(\int e^{\theta L_\theta f} d\mu\right)^{1/\theta} \leq 1. \tag{6.19}$$

Now, as a consequence of the convexity assumption on U,

$$\liminf_{\theta \to 1} L_\theta(x, y) \geq \frac{c}{2} |x - y|^2.$$

As a result, letting $\theta \to 1$ in (6.19), we arrive at

$$\int e^{Q_c f} d\mu \leq e^{\int f d\mu}$$

which thus holds true for all bounded measurable functions f. As a consequence of Corollary 6.4, the theorem is established. □

Although we need not really be concerned with that, it is important to emphasize that quadratic transportation cost inequalities are actually consequences of logarithmic Sobolev inequalities. In particular, a statement such as Theorem 6.5 above is a consequence of Theorem 5.2 for logarithmic Sobolev inequalities. The derivation of a quadratic transportation cost inequality from a logarithmic Sobolev inequality may be performed on the model of the Herbst argument. We briefly describe the argument. Assume thus we are given an absolutely continuous probability measure μ on the Borel sets of $\mathbb{R}^n$ such that for all smooth functions f on $\mathbb{R}^n$,

$$\mathrm{Ent}_\mu(f^2) \leq 2C \int |\nabla f|^2 d\mu. \tag{6.20}$$

Given a (bounded Lipschitz) function f on $\mathbb{R}^n$, apply now the logarithmic Sobolev inequality (6.20) to $e^{Q_{1/C}(\lambda f)/2}$, $\lambda \geq 0$, where we recall that $Q_c f$, $c > 0$, is the infimum-convolution of f with the quadratic cost $\tilde{c}(x-y) = \frac{c}{2}\,|x-y|^2$, $x, y \in \mathbb{R}^n$, $c > 0$. Now infimum-convolutions

$$v = v(x,t) = \inf_{y \in \mathbb{R}^n} \Big[f(y) + \frac{1}{2t}\,|x-y|^2 \Big], \quad x \in \mathbb{R}^n,\ t > 0,$$

are the Hopf-Lax representation of solutions of the Hamilton-Jacobi initial value problem

$$\begin{cases} \dfrac{\partial v}{\partial t} + \dfrac{1}{2}\,|\nabla v|^2 = 0 & \text{in } \mathbb{R}^n \times (0,\infty), \\ \qquad\qquad v = f & \text{on } \mathbb{R}^n \times \{t = 0\} \end{cases}$$

(cf. [Ev]). Therefore, setting

$$g = g(x,\lambda) = Q_{1/C}(\lambda f)(x),$$

almost everywhere in space,

$$g = \lambda \frac{\partial}{\partial \lambda}\, g + \frac{1}{2}\,|\nabla g|^2.$$

We thus immediately deduce from the logarithmic Sobolev inequality (6.20) the differential inequality

$$\lambda M'(\lambda) \leq M(\lambda) \log M(\lambda), \quad \lambda \geq 0,$$

on $M(\lambda) = \int e^g d\mu$. Since $M'(0) = \int f d\mu$, it follows as in the proof of Theorem 5.3 that

$$\int e^{Q_{1/C} f} d\mu \leq e^{\int f d\mu},$$

which is the infimum-convolution inequality (6.15).

As a consequence of the preceding and Corollary 6.4, we may state the following result.

Theorem 6.6. *Assume that μ is absolutely continuous and that for some $C > 0$ and all smooth enough functions f on $\mathbb{R}^n$,*

$$\mathrm{Ent}_\mu(f^2) \leq 2C \int |\nabla f|^2 d\mu.$$

Then, for every probability measure ν,

$$\mathrm{W}_2(\mu, \nu) \leq \sqrt{C\,\mathrm{H}(\nu\,|\,\mu)}.$$

Replacing $|x-y|$ by the Riemannian distance $d(x,y)$ would yield the same conclusion on a smooth complete Riemannian manifold (X, g). In particular, most of the logarithmic Sobolev inequalities we discussed in Chapter 5 may be turned into a quadratic transportation inequality. It may easily be shown (cf. [O-V], [B-G-L]) that the quadratic transportation cost inequality (6.14), or its equivalent infimum-convolution inequality (6.15), implies a Poincaré inequality (with constant C) so that quadratic transportation cost inequalities appear to be intermediate between logarithmic Sobolev inequalities and Poincaré inequalities (however producing normal concentration).

This line of reasoning may be pushed further to show that the modified logarithmic Sobolev inequality satisfied by the exponential distribution (Theorem 5.12) also implies a transportation cost inequality for the cost (4.20). For the particular example of the exponential distribution ν itself with density $\frac{1}{2}\,\mathrm{e}^{-|x|}$ with respect to Lebesgue measure on $\mathbb{R}$, we have the following statement.

Theorem 6.7. *For every probability measure ζ absolutely continuous with respect to the exponential distribution ν on $\mathbb{R}$,*

$$\mathrm{W}_{\tilde{c}}(\nu, \zeta) \leq C\,\mathrm{H}(\zeta\,|\,\nu)$$

where $\tilde{c}(x,y) = \tilde{c}(x-y)$ is the cost function (4.20) and $C > 0$ a universal constant.

The argument is quite similar to the one for the quadratic cost and we refer to [B-G-L] for details. The transportation cost inequality of Theorem 6.7 is equivalent to the one put forward in [Tal10]. It may be tensorized to dimension free transportation inequalities for products of the exponential distribution to recover its sharp concentration properties [Tal10] as exposed in Sections 4.5 and 5.3 by other means.

6.3 Transportation for product and non-product measures

In this section, we show how the transportation approach may be developed to obtain some of the geometric results of Sections 4.2 and 4.3. The arguments rely on somewhat delicate coupling arguments. One main interest in this approach is that it allows us to investigate some non-product situations as emphasized by K. Marton [Mar2], [Mar3]. We only review here a few recent results in this direction.

To avoid unessential measurability questions, for simplicity let (X_i, d_i), $i = 1, \ldots, n$, be arbitrary Polish (complete separable) metric spaces and let

$$X = X_1 \times \cdots \times X_n.$$

A point x in X has coordinates $x = (x_1, \ldots, x_n)$.

Denote by $\mathcal{P}(Q, R)$ the set of all probability measures on the product space $X \times X$ with marginals Q and R. Introduce a *coupling distance* between probability measures (cf. [R-R]) by

$$d(Q,R) = \inf \bigg(\int \sum_{i=1}^{n} \pi^y \big(\{x; x_i \neq y_i\} \big)^2 dR(y) \bigg)^{1/2}$$

where the infimum is running over all $\pi \in \mathcal{P}(Q, R)$. If π is a probability measure on a product space $\Sigma^k = \Sigma_1 \times \cdots \times \Sigma_k$, we denote by π^z, where $z = (z_{i_1}, \ldots, z_{i_\ell}) \in \Sigma^I = \Sigma_{i_1} \times \cdots \times \Sigma_{i_\ell}$, $I = \{i_1, \ldots, i_\ell\} \subset \{1, \ldots, k\}$, the (regular) conditional distribution of π given z, that is such that for every bounded measurable function ϕ on Σ^k,

$$\int \phi d\pi = \int_{\Sigma^I} \bigg(\int_{\Sigma^{\{1,\ldots,k\} \setminus I}} \phi(w, z) d\pi^z(w) \bigg) d\pi_I(z)$$

where π_I is the marginal of π over Σ^I (cf. e.g. [Str2]). In probabilistic notation,

$$d(Q,R) = \inf \bigg(\int \sum_{i=1}^{n} \mathbb{P}\big(\{\xi_i \neq \zeta_i \,|\, \zeta = y\} \big)^2 dR(y) \bigg)^{1/2} \tag{6.21}$$

where the couple of random variables (ξ, ζ) has distribution π and $\xi = (\xi_1, \ldots, \xi_n)$, $\zeta = (\zeta_1, \ldots, \zeta_n)$. Note that $d(Q, R)$ is not symmetric in Q, R.

To better understand the significance of the coupling distance d, let us relate it to the convex hull distance $\mathcal{D}_A^c$ of Section 4.2. To this task recall from (4.5) that, for a given subset A of the product space X and $x \in A$, we let

$$\mathcal{D}_A^c(x) = \inf_{y \in V_A(x)} |y|$$

where $V_A(x)$ is the convex hull of $(\mathbf{1}_{\{y_1 \neq x_1\}}, \ldots, \mathbf{1}_{\{y_n \neq x_n\}})$, $y \in A$, in $[0,1]^n$. It is easily seen that by the Cauchy-Schwarz inequality,

$$\mathcal{D}_A^c(x)^2 = \inf \sum_{i=1}^{n} \nu\big(\{y; y_i \neq x_i\} \big)^2$$

where the infimum is running over all probability measures on X such that $\nu(A) = 1$. As a consequence, for any Q supported on A and any R,

$$\int (\mathcal{D}_A^c)^2 dR \leq d(Q,R)^2. \tag{6.22}$$

The following result will contain the main conclusions of Section 4.2.

Theorem 6.8. *For any product measure P on $X = X_1 \times \cdots \times X_n$ and any probability measures Q and R on X,*

$$\frac{1}{4}\, d(Q,R)^2 \le \mathrm{H}\big(R\,|\,P\big) + \mathrm{H}\big(Q\,|\,P\big).$$

To apply Theorem 6.8, take $A \subset X$ and assume that $P(A) > 0$. Define first $Q(\cdot) = P(\cdot\,|A)$, for which $\mathrm{H}(Q\,|\,P) = \log \frac{1}{P(A)}$. Let R be such that

$$\frac{dR}{dP} = \frac{1}{Z}\, \mathrm{e}^{(\mathcal{D}_A^c)^2/4}$$

where Z is the normalization constant for which

$$\mathrm{H}\big(R\,|\,P\big) = \int \tfrac{1}{4}\, (\mathcal{D}_A^c)^2 dR - \log Z.$$

By (6.22),

$$\mathrm{H}\big(R\,|\,P\big) \le \frac{1}{4}\, d(Q,R)^2 - \log Z.$$

It then follows from Theorem 6.8 that

$$\log Z \le \mathrm{H}\big(Q\,|\,P\big) = \log \frac{1}{P(A)},$$

that is,

$$Z = \int \mathrm{e}^{(\mathcal{D}_A^c)^2/4} dP \le \frac{1}{P(A)},$$

which is the content of Theorem 4.6.

The proof of Theorem 6.8 uses coupling arguments. We indicate the sketch of the proof. We actually deal with the more general version of Theorem 4.11. To this task, given $\beta \ge 0$, recall the function

$$\tau_\beta(1-u) = \beta u \log u - (1+\beta u)\log\Big(\frac{1+\beta u}{1+\beta}\Big), \quad u \in [0,1].$$

Define, for every $\beta \ge 0$, the coupling distance

$$d_\beta(Q,R) = \inf \int \sum_{i=1}^{n} \tau_\beta\big(\pi^y(\{x; x_i \ne y_i\})\big)dR(y)$$

where the infimum is taken as in (6.21). Note that $\tau_\beta(u) \ge \frac{u^2}{4}$ for $\beta = 1$. The following theorem thus covers Theorem 6.8 and will imply exactly in the same way Theorem 4.11.

Theorem 6.9. *For any $\beta \ge 0$, any product probability measure P on $X = X_1 \times \cdots \times X_n$ and any probability measures Q and R on X,*

$$d_\beta(Q,R) \le \mathrm{H}\big(R\,|\,P\big) + \beta\,\mathrm{H}\big(Q\,|\,P\big).$$

Proof. The following (one-dimensional) coupling lemma is the key to the proof of Theorem 6.9.

Lemma 6.10. *Let (X, d) be a Polish space. For any $\beta \geq 0$ and any probability measures Q and R on X, define*

$$\Delta_\beta(Q, R) = \int \tau_\beta\left(\left(1 - \frac{dQ}{dR}\right)^+\right) dR.$$

Then, there exists $\pi \in \mathcal{P}(Q, R)$ such that

$$\int \tau_\beta\big(\pi^y(\{x; x \neq y\}\big) dR(y) = \Delta_\beta(Q, R).$$

Furthermore, for any probability measure P on X,

$$\Delta_\beta(Q, R) \leq \mathrm{H}\big(R \,|\, P\big) + \beta\, \mathrm{H}\big(Q \,|\, P\big).$$

Proof. Let $(R - Q)^+$ denote the positive part of the finite signed measure $R - Q$ and let $Q \wedge R$ denote the positive measure

$$R - (R - Q)^+ = Q - (Q - R)^+.$$

Set $p = (R - Q)^+(X)$. Suppose $0 < p < 1$, and consider independent random variables U, V and W with respective distributions

$$(1 - p)^{-1} Q \wedge R, \quad p^{-1}(R - Q)^+ \quad \text{and} \quad p^{-1}(Q - R)^+.$$

Let $I \in \{1, 2\}$ be chosen, independently of these variables, such that $I = 2$ with probability p. Set $\xi = \zeta = U$ when $I = 1$ and $\xi = W \neq V = \zeta$ when $I = 2$. If $p = 1$, then we do not need the variable V for the construction of ξ, ζ, whereas for $p = 0$ we never use W and V. The coupling (ξ, ζ) with distribution $\pi \in \mathcal{P}(Q, R)$ is such that

$$\pi\big(\{(x, y); x \neq y, y \in \cdot\}\big) = (R - Q)^+(\cdot).$$

By definition of conditional probability distributions, for every Borel set B in X,

$$\begin{aligned}
\int_B \pi^y\big(\{x; x \neq y\}\big) dR(y) &= \int_B \pi^y\big(\{(x, y); x \neq y\}\big) dR(y) \\
&= \pi\Big(\bigcup_{y \in B} \{(x, y); x \neq y\}\Big) \\
&= (R - Q)^+(B)
\end{aligned}$$

implying that $\pi^y(\{x; x \neq y\}) = \big(1 - \frac{dQ}{dR}\big)^+(y)$ for R-almost every y. The first part of the lemma follows.

Turning to the second part, it suffices to consider P such that both $f = dR/dP$ and $g = dQ/dP$ exist. Let

$$P_\beta = \frac{1}{1 + \beta}(R + \beta Q)$$

so that

$$\frac{dP_\beta}{dP} = h = \frac{1}{1+\beta}\,(f+\beta g).$$

Hence

$$\begin{aligned}\mathrm{H}(R\,|\,P) + \beta\,\mathrm{H}(Q\,|\,P) &= \int \big[f\log f + \beta g\log g\big]dP \\ &\geq \int \Big[f\log\frac{f}{h} + \beta g\log\frac{g}{h}\Big]dP\end{aligned}$$

since $\int h\log h dP = H(P_\beta\,|\,P) \geq 0$. Set $\rho = dQ/dR$ so that

$$\frac{h}{f} = \frac{1+\beta\rho}{1+\beta} \quad \text{and} \quad \frac{g}{f} = \rho.$$

Hence, by the preceding,

$$\mathrm{H}(R\,|\,P) + \beta\,\mathrm{H}(Q\,|\,P) \geq \int \tau_\beta(1-\rho)dR$$

from which the desired claim follows since $\tau_\beta(1-u) \geq \tau_\beta((1-u)^+)$ for every $u \geq 0$. Lemma 6.10 is established. □

The proof of Theorem 6.9 is now the proper tensorization of Lemma 6.10.

Proof of Theorem 6.9. Fix $\beta > 0$ and $P = \mu_1 \otimes \cdots \otimes \mu_n$. For $i = 1, \dots, n$, denote by $Q^{x_1,\dots,x_{i-1}}$ the (regular) conditional distribution of Q given $x_1, \dots, x_{i-1}$, and by $Q_i^{x_1,\dots,x_{i-1}}$ its marginal along the x_i coordinate. Define $R_i^{x_1,\dots,x_{i-1}}$ similarly. By the first part of Lemma 6.10, for every $i = 1, \dots, n$, there exists a probability measure π_i in $\mathcal{P}(Q_i^{x_1,\dots,x_{i-1}}, R_i^{x_1,\dots,x_{i-1}})$ such that

$$\Delta_\beta\big(Q_i^{x_1,\dots,x_{i-1}}, R_i^{x_1,\dots,x_{i-1}}\big) = \int \tau_\beta\big(\pi_i^y(\{x; x \neq y\})\big)dR_i^{x_1,\dots,x_{i-1}}(y).$$

Now let π on $X \times X$ be defined by

$$\pi(B) = \int\cdots\int_B \pi_1(dx_1, dy_1)\pi_2^{x_1,y_1}(dx_2, dy_2)\cdots\pi_n^{x_1,\dots,x_{n-1},y_1,\dots,y_{n-1}}(dx_n, dy_n).$$

Note that $\pi \in \mathcal{P}(Q, R)$. By the properties of conditional expectations and convexity of τ_β, for every $i = 1, \dots, n$,

$$\begin{aligned}&\int \tau_\beta\big(\pi^y(\{x; x_i \neq y_i\})\big)d\pi(x,y) \\ &\qquad\leq \int \tau_\beta\big(\pi_i^{y_i}(\{x_i \neq y_i\})\big)d\pi(x,y) \\ &\qquad= \int \Delta_\beta\big(Q_i^{x_1,\dots,x_{i-1}}, R_i^{y_1,\dots,y_{i-1}}\big)d\pi(x,y).\end{aligned}$$

As a consequence of Lemma 6.10,

$$\begin{aligned}d_\beta(Q,R) &\leq \sum_{i=1}^n \int \Delta_\beta\big(Q_i^{x_1,\dots,x_{i-1}}, R_i^{y_1,\dots,y_{i-1}}\big)d\pi(x,y) \\ &\leq \int \sum_{i=1}^n \Big(\mathrm{H}\big(R_i^{y_1,\dots,y_{i-1}}\,|\,\mu_i\big) + \beta\,\mathrm{H}\big(Q_i^{x_1,\dots,x_{i-1}}\,|\,\mu_i\big)\Big)d\pi(x,y).\end{aligned}$$

The chain rule for relative entropy (cf. [De-S], [Dem-Z1]) indicates that

$$\mathrm{H}(Q\,|P) = \sum_{i=1}^{n} \int \mathrm{H}(Q_i^{x_1,\dots,x_{i-1}} \,|\, \mu_i)\,dQ(x)$$

and similarly for $\mathrm{H}(R\,|\,P)$. Theorem 6.9 is thus established. □

As shown by A. Dembo [De], the preceding tools may be extended to reach in the same way the q point approximation theorem of Section 4.3.

The transportation approach may be well suited to reach some dependent situations. Variants of Theorem 6.8 have been used by K. Marton [Mar2] in the investigation of some non-product Markov chains for which it appears to be a powerful tool. More precisely, let P be a Markov chain on $X = X_1 \times \cdots \times X_n$ with transition kernels Π_i, $i = 1, \dots, n$, that is,

$$dP(x_1, \dots, x_n) = \Pi_n(x_n, dx_{n-1}) \cdots \Pi_2(x_2, dx_1)\Pi_1(dx_1).$$

Assume that, for some $0 \le \rho < 1$, for every $i = 1, \dots, n$, and every $x, y \in [0,1]$,

$$\big\|\Pi_i(x,\cdot) - \Pi_i(y,\cdot)\big\|_{\mathrm{TV}} \le \rho. \tag{6.23}$$

The case $\rho = 0$ of course corresponds to independent kernels Π_i.

In a statement analogous to Theorem 6.8, K. Marton [Mar2] obtains, through coupling characterizations of the total variation distance and Pinsker type inequalities, the following result. Recall the quadratic coupling distance,

$$d(Q,R) = \inf \bigg(\int \sum_{i=1}^{n} \pi^y\big(\{x; x_i \neq y_i\}\big)^2 dR(y)\bigg)^{1/2}.$$

Theorem 6.11. *Let P be a Markov chain on $X = X_1 \times \cdots \times X_n$ satisfying (6.23) for some $0 \le \rho < 1$. For any probability measures Q and R on X,*

$$\frac{1}{4}\,(1-\rho)^2 d(Q,R)^2 \le \mathrm{H}(R\,|P) + \mathrm{H}(R\,|P).$$

In the same way we deduced Theorem 4.6 from Theorem 6.8, we get the following corollary.

Corollary 6.12. *Let P be a Markov chain on $X = X_1 \times \cdots \times X_n$ satisfying (6.23) for some $0 \le \rho < 1$. Then, for any measurable non-empty subset A of X,*

$$\int \mathrm{e}^{(1-\rho)^2(\mathcal{D}_A^c)^2/4} dP \le \frac{1}{P(A)}\,.$$

The next corollary appears as the proper extension of Corollary 4.10.

Corollary 6.13. *Let P be a Markov chain on $[0,1]^n$ satisfying (6.23) for some $0 \le \rho < 1$. For every convex 1-Lipschitz map F on $\mathbb{R}^n$ and any $r \ge 0$,*

$$P\big(\{|F - \textstyle\int F dP| \ge r\}\big) \le 4\,\mathrm{e}^{-(1-\rho)^2 r^2/4}.$$

These results have been extended in [Mar4], and independently in [Sa], to larger classes of dependent processes, including Doeblin recurrent Markov chains and Φ-mixing processes. We refer to [Mar4], [Sa] for details.

Notes and Remarks

The interest of information inequalities for the concentration of measure phenomenon was emphasized in a series of papers by K. Marton [Mar1], [Mar2]. Proposition 6.1 is due to S. Bobkov and F. Götze [B-G].

General references on mass transportation problems and minimal metrics are [Ra2], [R-R]. The quadratic transportation cost inequality for Gaussian measures is due to M. Talagrand [Tal10], with a proof that is further extended to strictly log-concave measures in [Bl1] by means of the Brenier-McCann mass transference theorem [Bre], [MC]. A simple transparent proof is provided in [CE]. Its infimum-convolution description is emphasized in [B-G], and further analyzed in [Bo-L3] and [B-G-L]. That quadratic transportation cost inequalities follow from logarithmic Sobolev inequalities (Theorem 6.6) is due to F. Otto and C. Villani [O-V] with a PDE proof. The connection with Hamilton-Jacobi equations and hypercontractivity is made clear in [B-G-L] (see also [An]). Transportation cost inequalities for the exponential distribution (Theorem 6.7) were first obtained in [Tal10] to produce an alternative approach to the concentration results of [Tal3] presented here in Section 4.5. The connection with modified logarithmic Sobolev inequalities is exposed similarly in [B-G-L].

The transportation cost approach was initiated by K. Marton [Mar2] to extend Talagrand's convex hull theorem to some contractive Markov chain. Theorem 6.11 is due to K. Marton [Mar2]. The method was then systematically applied by A. Dembo [De] to recover most of the geometric inequalities for product measures of Section 4.2 and 4.3. See also [Dem-Z2]. Theorem 6.9 and its proof are taken from [De] and [Dem-Z1]. More recent developments for dependent variables and processes are due to K. Marton [Mar3], [Mar4], P.-M. Samson [Sa] and E. Rio [Ri1] with applications to bounds on empirical processes in the spirit of the inequalities described in the next chapter.

7. SHARP BOUNDS ON GAUSSIAN AND EMPIRICAL PROCESSES

In this chapter, we illustrate some of the basic principles of concentration to sharp bounds on Gaussian and empirical processes (or norms of sums of independent random vectors). In the first section, we present the sharp deviation inequality for supremum of Gaussian processes and its consequence to strong integrability. We then turn to bounds on sums of independent random vectors and empirical processes, first with the geometric tools of Chapter 4. In the last section, we establish sharper bounds with the entropic method. One of the powers of the concentration inequalities is that they extend classical results for sums of samples of random variables to Lipschitz functions of such samples, therefore allowing new, powerful applications.

7.1 Gaussian processes

We illustrate in this section the application of concentration properties of Gaussian measures to sharp deviation and integrability results for Gaussian processes.

On some probability space $(\Omega, \mathcal{A}, \mathbb{P})$, let $G = (G_t)_{t\in T}$ be a centered Gaussian process indexed by some parameter set T. We mean thus that for any finite collection $(t_1, \ldots, t_n)$ in T, the random vector $(G_{t_1}, \ldots, G_{t_n})$ is a centered Gaussian random vector in $\mathbb{R}^n$.

We are interested here in sharp probability inequalities on the distribution of the supremum of G. To this task, assume, to avoid measurability questions, that T is countable. In general this is dispensed with using separability assumptions. In any case, the basic inequalities are finite dimensional. We assume that $\sup_{t\in T} G_t < \infty$ almost surely.

We first claim that

$$\sigma = \sup_{t\in T} \big(\mathbb{E}(G_t^2)\big)^{1/2}. \tag{7.1}$$

Indeed, let m be such that $\mathbb{P}(\{\sup_{t\in T} G_t \leq m\}) \geq \frac{3}{4}$. Then, for every t, we have $\mathbb{P}(\{G_t \leq m\}) \geq \frac{3}{4}$ and, if $\sigma_t = (\mathbb{E}(G_t^2))^{1/2}$, $\frac{m}{\sigma_t} \geq \Phi^{-1}(\frac{3}{4}) > 0$, from which (7.1) follows (recall that Φ is the distribution function of the standard one-dimensional Gaussian law).

Now fix $t_1, \ldots, t_n$ in T and consider the centered Gaussian random vector $(G_{t_1}, \ldots, G_{t_n})$ in $\mathbb{R}^n$. Denote by $\Gamma = \Xi\, {}^t\Xi$ its (semi-) positive definite covariance matrix. The random vector $(G_{t_1}, \ldots, G_{t_n})$ has thus the same distribution as $\Xi\mathcal{N}$ where $\mathcal{N}$ is distributed according to the canonical Gaussian measure γ on $\mathbb{R}^n$. Let

$F : \mathbb{R}^n \to \mathbb{R}$ be defined by

$$F(x) = \max_{1 \le i \le n} (\Xi x)_i, \quad x = (x_1, \dots, x_n) \in \mathbb{R}^n.$$

Hence the distribution of F under γ is the distribution of the random variable $\max_{1 \le i \le n} G_{t_i}$. It is easily seen that the Lipschitz norm $\|F\|_{\mathrm{Lip}}$ of F is less than or equal to the norm $\|\Xi\|$ of Ξ as an operator from $\mathbb{R}^n$ equipped with the Euclidean norm into $\mathbb{R}^n$ equipped with the supremum norm. Furthermore, this operator norm $\|\Xi\|$ is equal by construction to

$$\max_{1 \le i \le n} \Big(\sum_{j=1}^n \Xi_{ij}^2 \Big)^{1/2} = \max_{1 \le i \le n} \big(\mathbb{E}(G_{t_i}^2) \big)^{1/2}$$

where we denote by Ξ_{ij} the entries of the matrix Ξ. Applying thus (2.35) for example to F yields

$$\mathbb{P}\Big(\big\{ \max_{1 \le i \le n} G_{t_i} \ge \mathbb{E}\big(\max_{1 \le i \le n} G_{t_i} \big) + r \big\} \Big) \le \mathrm{e}^{-r^2/2\sigma^2} \tag{7.2}$$

for every $r \ge 0$. We then argue as in the proof of Proposition 1.7. The same inequality applied to $-F$ yields

$$\mathbb{P}\Big(\big\{ \max_{1 \le i \le n} G_{t_i} \le \mathbb{E}\big(\max_{1 \le i \le n} G_{t_i} \big) - r \big\} \Big) \le \mathrm{e}^{-r^2/2\sigma^2}. \tag{7.3}$$

Let r_0 be large enough so that $\mathrm{e}^{-r_0^2/2\sigma^2} < \frac{1}{2}$. Also let m be large enough in order that $\mathbb{P}(\{\sup_{t \in T} G_t \le m\}) \ge \frac{1}{2}$. Intersecting this probability with the one in (7.3) for $r = r_0$ shows that

$$\mathbb{E}\big(\max_{1 \le i \le n} G_{t_i} \big) \le r_0 + m.$$

Since m and r_0 have been chosen independently of $t_1, \dots, t_n$, we already get that

$$\mathbb{E}\big(\sup_{t \in T} G_t \big) < \infty.$$

Now, one can use monotone convergence in (7.2) to come to the following basic inequality.

Theorem 7.1. *Let $G = (G_t)_{t \in T}$ be a centered Gaussian process indexed by a countable set T such that $\sup_{t \in T} G_t < \infty$ almost surely. Then, $\mathbb{E}(\sup_{t \in T} G_t) < \infty$ and for every $r \ge 0$,*

$$\mathbb{P}\Big(\big\{ \sup_{t \in T} G_t \ge \mathbb{E}\big(\sup_{t \in T} G_t \big) + r \big\} \Big) \le \mathrm{e}^{-r^2/2\sigma^2}$$

where $\sigma^2 = \sup_{t \in T} \mathbb{E}(G_t^2) < \infty$.

Therefore, up to the deviation factor $\mathbb{E}(\sup_{t \in T} G_t)$, the distribution of the supremum $\sup_{t \in T} G_t$ is as good as a one-dimensional Gaussian law with variance the supremum of the variances. Furthermore, for every $r \ge 0$,

$$\mathbb{P}\Big(\big\{ \big| \sup_{t \in T} G_t - \mathbb{E}\big(\sup_{t \in T} G_t \big) \big| \ge r \big\} \Big) \le 2\,\mathrm{e}^{-r^2/2\sigma^2} \tag{7.4}$$

and

$$\operatorname{Var}\big(\sup_{t\in T} G_t\big) \le 4\sigma^2.$$

Both Theorem 7.1 and (7.4) hold similarly with the median of $\sup_{t\in T} G_t$ instead of the mean (using (2.10)). One may also work with $\sup_{t\in T}|G_t|$ assuming $\sup_{t\in T}|G_t| < \infty$ almost surely.

As we already discussed it in Chapter 3, the main interest in the inequality of Theorem 7.1 is the relative size of σ and $\mathbb{E}(\sup_{t\in T} G_t)$. We always have

$$\begin{aligned}
\mathbb{E}\big(\sup_{t\in T} G_t\big) &= \frac{1}{2}\,\mathbb{E}\big(\sup_{s,t\in T}(G_s - G_t)\big) \\
&= \frac{1}{2}\,\mathbb{E}\big(\sup_{s,t\in T}|G_s - G_t|\big) \\
&\ge \frac{1}{2}\sup_{s,t\in T}\Big(\frac{2}{\pi}\Big)^{1/2}\Big(\mathbb{E}\big(|G_s - G_t|^2\big)\Big)^{1/2}
\end{aligned}$$

so that, for any $s \in T$,

$$\sigma \le \big(\mathbb{E}(G_s^2)\big)^{1/2} + \sqrt{2\pi}\,\mathbb{E}\big(\sup_{t\in T} G_t\big).$$

In general $\mathbb{E}(\sup_{t\in T} G_t)$ is much bigger than σ. Think for example of G being distributed as γ on $\mathbb{R}^n$ for which $\mathbb{E}(\sup_{t\in T} G_t)$ is of the order of $\sqrt{\log n}$ for n large whereas $\sigma = 1$. This was actually one crucial point in the concentration proof of Dvoretzky's theorem in Section 3.5. On the other hand, $\mathbb{E}(\sup_{t\in T} G_t)$ only appears as a deviation factor and not as a multiplicative factor in the exponential. The concentration inequalities however do not provide any hint on the size of $\mathbb{E}(\sup_{t\in T} G_t)$ itself for which independent tools are required (entropy or majorizing measures (cf. [Le-T], [Tal11]).

The next theorem is a consequence of the strong integrability of the supremum of Gaussian processes.

Corollary 7.2. *Let $G = (G_t)_{t\in T}$ be a centered Gaussian process indexed by a countable set T such that $\sup_{t\in T} G_t < \infty$ almost surely. Then*

$$\lim_{r\to\infty}\frac{1}{r^2}\log\mathbb{P}\big(\big\{\sup_{t\in T} G_t \ge r\big\}\big) = -\frac{1}{2\sigma^2}$$

where $\sigma^2 = \sup_{t\in T}\mathbb{E}(G_t^2) < \infty$. Equivalently,

$$\mathbb{E}\Big(\exp\big(\rho(\sup_{t\in T} G_t)^2\big)\Big) < \infty$$

if and only if $\rho < \frac{1}{2\sigma^2}$.

The first assertion in Corollary 7.2 is a large deviation statement for complements of balls. The upper bound immediately follows from Theorem 7.1. The lower bound is just that

$$\begin{aligned}
\mathbb{P}\big(\big\{\sup_{t\in T} G_t \ge r\big\}\big) &\ge \mathbb{P}\big(\{G_t \ge r\}\big) \\
&= 1 - \Phi\Big(\frac{r}{\sigma_t}\Big) \\
&\ge \frac{\mathrm{e}^{-r^2/2\sigma_t^2}}{\sqrt{2\pi}(1 + (r/\sigma_t))}
\end{aligned}$$

for every $t \in T$ and $r \geq 0$. The second part of the theorem follows by integration of the inequality of Theorem 7.1 in $r \geq 0$.

The deviation inequality of Theorem 7.1 actually contains much more information than the integrability result of Corollary 7.2. For example, if G^n is a sequence of Gaussian processes as before, and if we let $\|G^n\| = \sup_{t\in T} G_t^n$, $n \in \mathbb{N}$, then $\|G^n\| \to 0$ almost surely as soon as $\mathbb{E}(\|G^n\|) \to 0$ and $\sigma^n\sqrt{\log n} \to 0$ where, for every n, $\sigma^n = \sup_{t\in T}(\mathbb{E}((G_t^n)^2))^{1/2}$.

As we have seen above, $\mathbb{E}\big(\sup_{t\in T}|G_t|\big) \geq \kappa\sigma$ for some numerical $\kappa > 0$. Therefore, integration of (7.4) applied to $\sup_{t\in T}|G_t|$ shows, as in Proposition 1.10 and (1.19), that for some numerical constant $C > 0$ and all $q \geq 1$,

$$\Big(\mathbb{E}\big(\sup_{t\in T}|G_t|^q\big)\Big)^{1/q} \leq C\sqrt{q}\,\mathbb{E}\big(\sup_{t\in T}|G_t|\big). \tag{7.5}$$

This equivalence of moments is a useful tool in the study of Gaussian processes and measures.

The preceding results may be presented equivalently on Gaussian measures on infinite dimensional normed vector spaces. Now, such Gaussian measures may be represented as (almost sure convergent) series $\sum_{i=1}^\infty g_i v_i$ where v_i, $i \geq 1$, is a sequence in $(E, \|\cdot\|)$ and the g_i's are independent standard normal random variables on some probability space $(\Omega, \mathcal{A}, \mathbb{P})$. Theorem 7.1 and Corollary 7.2 thus describe the distribution and integrability properties of the norm

$$\Big\|\sum_{i=1}^\infty g_i v_i\Big\|.$$

Note that since

$$\Big\|\sum_{i=1}^\infty g_i v_i\Big\| = \sup_{\|\xi\|\leq 1}\sum_{i=1}^\infty g_i\langle\xi, v_i\rangle$$

where the supremum is running over the unit ball of the dual space,

$$\sigma^2 = \sup_{\|\xi\|\leq 1}\sum_{i=1}^\infty \langle\xi, v_i\rangle^2.$$

Since the norm is convex, the results in Section 4.2 extend these conclusions to the supremum of functionals that are not necessarily Gaussian. For simplicity we deal with finite sums. Indeed, let η_i, $i = 1, \dots, n$, be independent random variables on $(\Omega, \mathcal{A}, \mathbb{P})$ with $|\eta_i| \leq 1$ almost surely. Let v_i, $i = 1, \dots, n$, be vectors in an arbitrary normed space E with norm $\|\cdot\|$. Since the norm is a supremum of linear functionals, we may directly apply Corollary 4.8 to get deviation and concentration inequalities around a median. Let us however repeat here the argument. Consider the convex function $F : \mathbb{R}^n \to \mathbb{R}$ defined by

$$F(x) = \Big\|\sum_{i=1}^n x_i v_i\Big\|, \quad x = (x_1, \dots, x_n) \in \mathbb{R}^n.$$

Then, by duality, for $x, y \in \mathbb{R}^n$,

$$\begin{aligned}\big|F(x) - F(x)\big| &\leq \Big\|\sum_{i=1}^n (x_i - y_i)v_i\Big\| \\ &= \sup_{\|\xi\|\leq 1}\sum_{i=1}^n (x_i - y_i)\langle\xi, v_i\rangle \leq \sigma|x - y|,\end{aligned}$$

where the last step is obtained from the Cauchy-Schwarz inequality. Hence $\|F\|_{\mathrm{Lip}} \leq \sigma$. As an application of Corollary 4.10, or Theorem 5.9, we get the following result.

Theorem 7.3. *Let $\eta_1, \ldots, \eta_n$ be independent random variables such that $|\eta_i| \leq 1$ almost surely, $i = 1, \ldots, n$, and let $v_1, \ldots, v_n$ be vectors in a normed space $(E, \|\cdot\|)$. For every $r \geq 0$,*

$$\mathbb{P}\left(\left\{\left\|\sum_{i=1}^{n} \eta_i v_i\right\| \geq M + r\right\}\right) \leq 2\,\mathrm{e}^{-r^2/16\sigma^2}$$

where M is either the mean or a median of $\|\sum_{i=1}^{n} \eta_i v_i\|$ and where

$$\sigma^2 = \sup_{\|\xi\| \leq 1} \sum_{i=1}^{n} \langle \xi, v_i \rangle^2.$$

There is a similar inequality for deviations under M, and thus a concentration inequality.

This inequality is the analogue of the Gaussian deviation inequality of Theorem 7.1 and describes one proper infinite dimensional extension of the Hoeffding type inequality (1.23).

Theorem 7.3 may be used as for Corollary 7.2 to show that whenever the series

$$S = \sum_{i=1}^{\infty} \eta_i v_i$$

is almost surely convergent in E, then its norm $\|S\|$ is strongly integrable in the sense that

$$\mathbb{E}\big(\mathrm{e}^{\rho\|S\|^2}\big) < \infty \tag{7.6}$$

for every ρ. (The fact that it holds for every ρ with respect to the Gaussian case is due to the fact that the η_i's are bounded.)

Theorem 7.3 may be used to prove equivalences of moments as for Gaussian random vectors and series. In particular, if the η_i's are symmetric Bernoulli variables taking values ± 1, the classical Khintchine inequalities show that

$$\begin{aligned}\mathbb{E}\big(\|S\|\big) &\geq \sup_{\|\xi\|\leq 1} \mathbb{E}\big(|\langle \xi, S\rangle|\big) \\ &\geq \sup_{\|\xi\|\leq 1} \kappa\Big(\mathbb{E}\big(|\langle \xi, S\rangle|^2\big)\Big)^{1/2} = \kappa\sigma\end{aligned}$$

for some numerical $\kappa > 0$. Hence, as for (7.5),

$$\Big(\mathbb{E}\big(\|S\|^q\big)\Big)^{1/q} \leq C\sqrt{q}\,\mathbb{E}\big(\|S\|\big) \tag{7.7}$$

for some numerical constant $C > 0$ and all $q \geq 1$. Inequalities (7.7) are the famous Khintchine-Kahane inequalities. As for (7.5), these moment equivalences are part of the geometric reversed Hölder inequalities (2.21) and improve upon the earlier version (2.22) with the optimal growth of the constants in $q \geq 1$ (cf. [Le-T]).

7.2 Bounds on empirical processes

Sums of independent random variables are a natural application of the deviation inequalities for product measures. In this section and the next one, we present estimates on supremum of empirical processes. This paragraph relies on the geometric concentration properties developed in Sections 4.2 and 4.3 while the next one is based on the entropic method of Chapter 5.

Tail probabilities for sums of independent random variables have been extensively studied in classical probability theory and limit theorems. We already mentioned in this work the Hoeffding type inequality (1.23) (and its martingale extension Lemma 4.1)

$$\mathbb{P}\big(\{S \geq \mathbb{E}(S) + r\}\big) \leq \mathrm{e}^{-r^2/2nC^2}$$

where $S = Y_1 + \cdots + Y_n$ is a sum of independent real random variables bounded by $C > 0$. One finished result is the so-called Bennett inequality [Benn] (after contributions by S. Bernstein, A. Kolmogorov, Y. Prokhorov, W. Hoeffding, etc.). As for (1.23), it will be convenient to compare the infinite dimensional extension we will present with this result. With due respect to the Hoeffding inequality, this result takes into account the fluctuations of S with respect to the variance more carefully. Namely, let $Y_1, \ldots, Y_n$ be independent real-valued random variables on some probability space $(\Omega, \mathcal{A}, \mathbb{P})$ such that $|Y_i| \leq C$, $i = 1, \ldots, n$. Set $S = Y_1 + \cdots + Y_n$. Then, for every $r \geq 0$,

$$\mathbb{P}\big(\{S \geq \mathbb{E}(S) + r\}\big) \leq \exp\Big(-\frac{\sigma^2}{C^2}\, h\Big(\frac{Cr}{\sigma^2}\Big)\Big) \tag{7.8}$$

where $h(u) = (1+u)\log(1+u) - u$, $u \geq 0$, and $\sigma^2 = \sum_{i=1}^n \mathbb{E}(Y_i^2)$. Such an inequality describes the Gaussian tail for the values of r which are small with respect to σ^2, and the Poissonian behavior for the large values (think for example of a sample of independent Bernoulli variables, with probability of success either $\frac{1}{2}$ or of the order of $\frac{1}{n}$).

Our task in this chapter will be to understand what is saved of such a sharp inequality for norms of sums $S = \sum_{i=1}^n Y_i$ of independent random vectors Y_i taking values in some Banach space $(E, \|\cdot\|)$. In particular these bounds aim to be as close as possible to the one-dimensional inequality (7.8). They should also compare to the bounds on Gaussian processes of the preceding section involving two main parameters, one on the supremum itself (mean or median), and one on supremum of weak variances.

A first result in this direction is Corollary 4.5 obtained from martingale inequalities. However, the inequality of Corollary 4.5, while a deviation inequality from the mean $\mathbb{E}(\|S\|)$, does not emphasize, according to Theorems 7.1 and 7.3, the supremum of weak variances

$$\sup_{\|\xi\| \leq 1} \sum_{i=1}^n \mathbb{E}\big(\langle \xi, Y_i\rangle^2\big).$$

This is why we have to turn to more refined tools such as the concentration inequalities in product spaces of Sections 4.2 and 4.3 or the entropic method of Chapter 5.

Motivated by recent applications, we present the results in the context of supremum of empirical processes rather than sums of random vectors. This is however only different at a notational level. In statistical applications, one is interested in such bounds uniformly over classes of functions, and importance of such inequalities has been emphasized in the statistical treatment of selection of models in [B-M1], [B-M2], [B-B-M] (see [Mas2]). More precisely, let $Y_1, Y_2, \dots, Y_n, \dots$ be independent random variables with values in some measurable space $(V, \mathcal{V})$ with identical distribution $\mathcal{P}$, and for $n \geq 1$, let

$$\mathcal{P}_n = \frac{1}{n} \sum_{i=1}^{n} \delta_{Y_i}$$

be the empirical measures (on $\mathcal{P}$). A class $\mathcal{F}$ of real measurable functions on V is said to be a Glivenko-Cantelli class if the supremum $\sup_{f \in \mathcal{F}} |\mathcal{P}_n(f) - \mathcal{P}(f)|$ converges almost surely to 0. It is a Donsker class if (in a sense to be made precise), $\sqrt{n}(\mathcal{P}_n(f) - \mathcal{P}(f))$, $f \in \mathcal{F}$, converges in distribution towards a centered Gaussian process $G_{\mathcal{P}} = \{G_{\mathcal{P}}(f),\ f \in \mathcal{F}\}$ with covariance function $\mathcal{P}(fg) - \mathcal{P}(f)\mathcal{P}(g)$, $f, g \in \mathcal{F}$. These definitions naturally extend the classical example of the class of all characteristic functions of intervals $(-\infty, t]$, $t \in \mathbb{R}$ (studied precisely by Glivenko-Cantelli and Donsker). These asymptotic properties however often need to be turned into tail inequalities at fixed n on classes $\mathcal{F}$ which are as rich as possible (to determine accurate approximation by empirical models). In particular, the aim is to reach exact extension of the Gaussian bound of Theorem 7.1 for $G_{\mathcal{P}}$ and the Bennett inequality (7.8) corresponding to a class $\mathcal{F}$ reduced to one function.

In this section, we present a result on the basis of the control by q points of Section 4.3. As we have seen there, this method is of particular interest to bounds on sums of independent random variables and we already presented there a useful inequality for non-negative summands (Corollary 4.14). In the study of empirical processes, one does not usually deal with non-negative summands. One general situation is thus the following. Let $Y_1, \dots, Y_n$ be independent random variables taking values in some space V and consider, say, a countable family $\mathcal{F}$ of (measurable) real-valued functions on V. We are interested in bounds on the tail of

$$Z = \sup_{f \in \mathcal{F}} \sum_{i=1}^{n} f(Y_i).$$

(One can deal similarly with absolute values

$$\sup_{f \in \mathcal{F}} \left| \sum_{i=1}^{n} f(Y_i) \right| .)$$

For simplicity, we deal with centered random variables, and thus assume that $\mathbb{E}(f(Y_i)) = 0$ for every $i = 1, \dots, n$ and every $f \in \mathcal{F}$. If this is not the case, replace $f(Y_i)$ by $f(Y_i) - \mathbb{E}(f(Y_i))$, although several results do actually still hold for $f(Y_i)$ itself. Standard symmetrization techniques reduce then to the investigation of

$$Z^s = \sup_{f \in \mathcal{F}} \sum_{i=1}^{n} \varepsilon_i f(Y_i)$$

where $\varepsilon_1, \dots, \varepsilon_n$ are independent symmetric Bernoulli random variables independent of the Y_i's. Indeed, if $\{Y_1', \dots, Y_n'\}$ denotes an independent copy of the sample $\{Y_1, \dots, Y_n\}$, it is not difficult to see that for every $r, s \geq 0$,

$$\mathbb{P}\big(\{Z \geq r + s\}\big) \leq \mathbb{P}\bigg(\bigg\{ \sup_{f \in \mathcal{F}} \sum_{i=1}^n \big(f(Y_i) - f(Y_i')\big) \geq r \bigg\}\bigg) + \sup_{f \in \mathcal{F}} \mathbb{P}\bigg(\bigg\{ \sum_{i=1}^n f(Y_i) \geq s \bigg\}\bigg).$$

By symmetry,

$$\sup_{f \in \mathcal{F}} \sum_{i=1}^n \big(f(Y_i) - f(Y_i')\big) \quad \text{and} \quad \sup_{f \in \mathcal{F}} \sum_{i=1}^n \varepsilon_i \big(f(Y_i) - f(Y_i')\big)$$

have the same distribution. Together with some minor modifications, we may thus reduce to the symmetrized supremum Z^s.

When dealing with Z or Z^s, we may not use directly Corollary 4.14. To overcome this difficulty, we use the symmetry properties of Z^s. Write

$$Z^s = \big(Z^s - \mathbb{E}_\varepsilon(Z^s)\big) + \mathbb{E}_\varepsilon(Z^s)$$

where $\mathbb{E}_\varepsilon$ is partial integration with respect to the Bernoulli variables $\varepsilon_1, \dots, \varepsilon_n$. The point is that, as we have seen prior to Corollary 4.13, $\mathbb{E}_\varepsilon(Z^s)$ is monotone and subadditive with respect to the independent random variables $Y_1, \dots, Y_n$. Corollary 4.13 therefore applies to $\mathbb{E}_\varepsilon(Z^s)$. The remainder term $Z^s - \mathbb{E}(Z^s)$ is bounded, conditionally on the Y_i's, with the deviation inequality of Theorem 7.3 by a Gaussian tail involving a random supremum of weak variances

$$\Sigma^2 = \sup_{f \in \mathcal{F}} \sum_{i=1}^n f^2(Y_i).$$

Now Σ^2 is again monotone and subadditive so that Corollary 4.13 may also be applied to it. Assuming $|f| \leq 1$, $f \in \mathcal{F}$, and combining the arguments yields after some work (cf. [Le-T], [Tal5]) the inequality, for integers $k, q \geq 1$,

$$\mathbb{P}\big(\big\{Z^s \geq 8q\mathbb{E}(Z^s) + 2k\big\}\big) \leq 2^q q^{-k} + 2\,\mathrm{e}^{-k^2/128q\mathbb{E}(\Sigma^2)}. \tag{7.9}$$

How is (7.9) used in applications? One possible choice is simply $q = 2$. We may also optimize the choice of q and take it of the order of $k/\log k$. This choice then leads to the following general statement (cf. [Tal5]).

Theorem 7.4. *If $|f| \leq C$ for every f in $\mathcal{F}$, and if $\mathbb{E}(f(Y_i)) = 0$ for every $f \in \mathcal{F}$ and $i = 1, \dots, n$, then, for all $r \geq 0$,*

$$\mathbb{P}\big(Z^s \geq K\mathbb{E}(Z^s) + r\big) \leq K \exp\bigg(-\frac{r}{KC} \log\bigg(1 + \frac{Cr}{\mathbb{E}(\Sigma^2) + C^2}\bigg)\bigg)$$

where $K > 0$ is a numerical constant.

To compare more carefully this result with (7.8), it is important to give a more tractable form to $\mathbb{E}(\Sigma^2)$. To this task, we may write

$$\mathbb{E}(\Sigma^2) \leq \sigma^2 + \mathbb{E}\Big(\sup_{f\in\mathcal{F}} \sum_{i=1}^n \big(f^2(Y_i) - \mathbb{E}(f^2(Y_i))\big)\Big).$$

By Jensen's inequality,

$$\mathbb{E}\Big(\sup_{f\in\mathcal{F}} \sum_{i=1}^n \big(f^2(Y_i) - \mathbb{E}(f^2(Y_i))\big)\Big) \leq \mathbb{E}\Big(\sup_{f\in\mathcal{F}} \sum_{i=1}^n \big(f^2(Y_i) - f^2(Y_i')\big)\Big)$$

where the Y_i''s are independent copies of the Y_i's, $i = 1, \ldots, n$. By symmetry,

$$\mathbb{E}\Big(\sup_{f\in\mathcal{F}} \sum_{i=1}^n \big(f^2(Y_i) - f^2(Y_i')\big)\Big) \leq 2\mathbb{E}\Big(\sup_{f\in\mathcal{F}} \Big|\sum_{i=1}^n \varepsilon_i f^2(Y_i)\Big|\Big)$$

where we recall that $\varepsilon_1, \ldots, \varepsilon_n$ are independent ± 1 symmetric Bernoulli random variables independent of the Y_i's. We then have to make use of a contraction inequality [Le-T], p. 112,

$$\mathbb{E}\Big(\sup_{f\in\mathcal{F}} \Big|\sum_{i=1}^n \varepsilon_i f^2(Y_i)\Big|\Big) \leq 4C\mathbb{E}\Big(\sup_{f\in\mathcal{F}} \Big|\sum_{i=1}^n \varepsilon_i f(Y_i)\Big|\Big).$$

Alltogether, we conclude that

$$\mathbb{E}(\Sigma^2) \leq \sigma^2 + 16C\mathbb{E}(\overline{Z}) \tag{7.10}$$

where

$$\sigma^2 = \sup_{f\in\mathcal{F}} \sum_{i=1}^n \mathbb{E}\big(f^2(Y_i)\big)$$

and

$$\overline{Z} = \sup_{f\in\mathcal{F}} \Big|\sum_{i=1}^n f(Y_i)\Big|$$

(if Z is defined without absolute values). With this observation, Theorem 7.4 provides an inequality close to both the classical exponential inequalities for sums of independent real-valued random variables (7.8) and to the bounds of the preceding section on supremum of Gaussian processes with the same basic parameters σ and $\mathbb{E}(Z)$. Theorem 7.4 is good enough to establish most basic almost sure limit theorems in Banach spaces (cf. [Le-T], [Le-Z]]).

7.3 Sharper bounds via the entropic method

One weak aspect of Theorem 7.4 is that it does not present a deviation inequality from the mean (or median) itself but rather a multiple of the mean. Neither does it provides a concentration inequality. This can be obtained, however, in a rather delicate way, by further abstract developments of the methods of Sections 4.2 and 4.3 as demonstrated in [Tal9].

Here, we observe that the functional approach based on logarithmic Sobolev inequalities of Chapter 5 may be used to yield these sharper bounds. The overall approach is simpler and more transparent than the preceding developments based on the results of Sections 4.2 and 4.3.

We start with a first result concerning supremum of empirical processes over a class of non-negative bounded functions. As before, let $Y_1, \dots, Y_n$ be a sample of independent random variables with values in some measurable space $(V, \mathcal{V})$. Let $\mathcal{F}$ be a (countable) class of measurable functions f on V such that $0 \leq f \leq 1$. Set

$$Z = \sup_{f \in \mathcal{F}} \sum_{i=1}^{n} f(Y_i).$$

Theorem 7.5. *Under the preceding notation, for any $\lambda \geq 0$,*

$$\mathbb{E}\big(e^{\lambda Z}\big) \leq e^{(e^{\lambda}-1)\mathbb{E}(Z)}.$$

In particular, for any $r \geq 0$,

$$\mathbb{P}\big(\{Z \geq \mathbb{E}(Z) + r\}\big) \leq \exp\bigg(-\mathbb{E}(Z)h\Big(\frac{r}{\mathbb{E}(Z)}\Big)\bigg)$$

where $h(u) = (1+u)\log(1+u) - u$, $u \geq 0$.

The exponential inequality of Theorem 7.5 is the optimal extension of (7.8) for non-negative summands. It is in particular attained when Z has a Poisson distribution. Note that since $h(u) \geq \frac{1}{2}\, u \log(1+u)$ for any $u \geq 0$, Theorem 7.5 implies that

$$\mathbb{P}\big(\{Z \geq \mathbb{E}(Z) + r\}\big) \leq \exp\bigg(-\frac{r}{2}\log\Big(1 + \frac{r}{\mathbb{E}(Z)}\Big)\bigg) \tag{7.11}$$

for every $r \geq 0$.

Proof. We may clearly assume that $\mathcal{F}$ is a finite class (with N elements). We may then represent Z as a function

$$Z(x) = \max_{1 \leq k \leq N} \sum_{i=1}^{n} x_i^k$$

where $x = (x_1, \dots, x_n) \in E^n$, $E = \mathbb{R}^N$. If we further denote by μ_i the law of $f(Y_i)$, $f \in \mathcal{F}$, on E, the distribution of Z under $\mathbb{P}$ is the same as the distribution of $Z(x)$ under the product measure P of the μ_i's. Note that since $0 \leq f \leq 1$, the μ_i's are supported by $[0,1]^N \subset E$, and we may thus assume that $E = [0,1]^N$.

Since P is a product measure, we may apply the product property of entropy (Proposition 5.6) to get that

$$\mathrm{Ent}_P(\mathrm{e}^{\lambda Z}) \leq \sum_{i=1}^{n} \int \mathrm{Ent}_{\mu_i}(\mathrm{e}^{\lambda Z_i}) dP. \tag{7.12}$$

Recall that here Z_i is Z as a function of the i-th variables, the others being fixed. Let $\phi(u) = \mathrm{e}^{-u} + u - 1$, $u \in \mathbb{R}$. The variational characterization of entropy (cf. (5.11)) indicates that for any probability measure μ and any (say bounded) function f,

$$\begin{aligned}\mathrm{Ent}_\mu(\mathrm{e}^f) &= \inf_{t \geq 0} \int \big[f\mathrm{e}^f - (\log t + 1)\,\mathrm{e}^f + t\big]\, d\mu \\ &= \inf_{u \in \mathbb{R}} \int \phi(f - u)\,\mathrm{e}^f\, d\mu.\end{aligned}$$

For every $x = (x_1, \dots, x_n)$ in E^n, and $i = 1, \dots, n$, set

$$y = y_i(x) = (x_1, \dots, x_{i-1}, 0, x_{i+1}, \dots, x_n) \in E^n.$$

In the variational characterization of $\mathrm{Ent}_{\mu_i}(\mathrm{e}^{\lambda Z_i})$, choose then $u = \lambda Z(y)$ so that

$$\int \mathrm{Ent}_{\mu_i}(\mathrm{e}^{\lambda Z_i}) dP \leq \int \phi\Big(\lambda\big(Z(x) - Z\big(y_i(x)\big)\big)\Big)\,\mathrm{e}^{\lambda Z(x)} dP(x). \tag{7.13}$$

Note that since $x_i \in [0,1]^N$, $Z(x) \geq Z(y_i(x))$ by definition of Z. On the other hand, if $(A_k)_{1 \leq k \leq N}$ is a partition of E^n such that

$$A_k \subset \Big\{x \in E^n; Z(x) = \sum_{i=1}^{n} x_i^k\Big\},$$

and if $\tau_k = \tau_k(x) = \mathbf{1}_{A_k}(x)$,

$$0 \leq Z(x) - Z\big(y_i(x)\big) \leq \sum_{k=1}^{N} \tau_k x_i^k = \tau \cdot x_i.$$

Now, since ϕ is convex, for any $\lambda \geq 0$ and $u \in [0,1]$, $\phi(\lambda u) \leq u\phi(\lambda)$. Therefore, for any $\lambda \geq 0$,

$$\phi\Big(\lambda\big(Z(x) - Z\big(y_i(x)\big)\big)\Big) \leq (\tau \cdot x_i)\,\phi(\lambda).$$

Together with (7.12) and (7.13), it follows that

$$\begin{aligned}\mathrm{Ent}_P(\mathrm{e}^{\lambda Z}) &\leq \phi(\lambda) \sum_{i=1}^{n} \int (\tau \cdot x_i)\,\mathrm{e}^{\lambda Z(x)} dP(x) \\ &= \phi(\lambda) \int Z\mathrm{e}^{\lambda Z} dP\end{aligned} \tag{7.14}$$

since by construction $\sum_{i=1}^{n} \tau \cdot x_i = Z(x)$.

Now let $\Lambda(\lambda) = \mathbb{E}(e^{\lambda Z}) = \int e^{\lambda Z} dP$, $\lambda \geq 0$, be the Laplace transform of Z. What we have obtained in (7.14) is that, for every $\lambda \geq 0$,

$$\lambda\Lambda'(\lambda) - \Lambda(\lambda)\log\Lambda(\lambda) \leq \phi(\lambda)\Lambda'(\lambda),$$

that is,

$$(1 - e^{-\lambda})\Lambda'(\lambda) \leq \Lambda(\lambda)\log\Lambda(\lambda), \quad \lambda \geq 0.$$

This differential inequality may be integrated in the optimal way. The function $J(\lambda) = \log\Lambda(\lambda)$ satisfies

$$J'(\lambda) \leq \big(1 - e^{-\lambda}\big)^{-1} J(\lambda),$$

that is,

$$\big[\log J(\lambda)\big]' \leq \big[\log(e^{\lambda} - 1)\big]'.$$

Hence $(e^{\lambda} - 1)J(\lambda) \nearrow \mathbb{E}(Z)$ as $\lambda \searrow 0$ from which the desired claim follows. By Chebyshev's inequality, for every $\lambda, r \geq 0$,

$$\mathbb{P}\big(\{Z \geq \mathbb{E}(Z) + r\}\big) \leq \mathbb{E}(Z)\phi(\lambda)\, e^{-\lambda r}$$

and minimizing over $\lambda \geq 0$ yields the deviation inequality of the statement. Theorem 7.5 is established. □

Now we turn to classes of (bounded) functions with arbitrary signs. As before, let $Y_1, \ldots, Y_n$ be independent random variables with values in some space V, and let $\mathcal{F}$ be a countable class of measurable functions on V. Set

$$Z = \sup_{f\in\mathcal{F}} \sum_{i=1}^{n} f(Y_i).$$

The arguments developed below are similar for

$$\overline{Z} = \sup_{f\in\mathcal{F}} \bigg|\sum_{i=1}^{n} f(Y_i)\bigg|.$$

Theorem 7.6. *If $|f| \leq C$ for every f in $\mathcal{F}$, then, for all $r \geq 0$,*

$$\mathbb{P}\Big(\big\{|Z - \mathbb{E}(Z)| \geq r\big\}\Big) \leq 3\exp\bigg(-\frac{r}{KC}\log\bigg(1 + \frac{Cr}{\mathbb{E}(\Sigma^2)}\bigg)\bigg)$$

where $\Sigma^2 = \sup_{f\in\mathcal{F}} \sum_{i=1}^{n} f^2(Y_i)$ is the supremum of random variances and where $K > 0$ is a numerical constant.

This statement is as close as possible to (7.8). With respect to Theorem 7.4, the main feature is the concentration property around the mean $\mathbb{E}(Z)$ (or equivalently a median of Z). As for Gaussian processes, the theorem does not yield any information on the size of $\mathbb{E}(Z)$ itself.

The proof of Theorem 7.6 relies, as the one of Theorem 7.5, on the entropic method but several technicalities make it a little heavier. In particular, the proof is made harder by the fact that we are looking for a concentration inequality.

Deviation above the mean (as for Theorem 5.9) requires significantly less effort, and even allows sharp bounds (cf. [Mas1], [Ri2], [Ri3]). For simplicity, we will not look that time for an optimal integration of the basic differential inequality on Laplace transforms we will get. Reasonable values of K are however deduced in [Mas1].

Proof. By homogeneity, take $C = 1$. As in the proof of Theorem 7.5, we may and do assume that $\mathcal{F}$ is finite. We then represent Z as a function

$$Z(x) = \max_{1 \le k \le N} \sum_{i=1}^{n} x_i^k$$

where $x = (x_1, \dots, x_n) \in E^n$, $E = \mathbb{R}^N$. As before, if we denote by μ_i the law of $f(Y_i)$, $f \in \mathcal{F}$, on E, the distribution of Z under $\mathbb{P}$ is the same as the distribution of $Z(x)$ under the product measure P of the μ_i's. With respect to Theorem 7.5, since $|f| \le 1$, the μ_i's are rather supported by $[-1, +1]^N \subset E$ so that we consider this time that $E = [-1, +1]^N$. Recall also that we denote by $(A_k)_{1 \le k \le N}$ a partition of E^n such that

$$A_k \subset \left\{ x \in E^n; Z(x) = \sum_{i=1}^{n} x_i^k \right\}.$$

Now, the convexity properties of Z ensure that for every $x \in E^n$, every $i = 1, \dots, n$ and every y such that $y_j = x_j$, $j \ne i$,

$$Z(x) - Z(y) \le \sum_{k=1}^{N} \tau_k |x_i^k - y_i^k| \tag{7.15}$$

where we recall that $\tau_k = \tau_k(x) = \mathbf{1}_{A_k}(x)$. On the other hand, since $x_i, y_i \in [-1, +1]^N$,

$$\big| Z(x) - Z(y) \big| \le 2. \tag{7.16}$$

We make use of Corollary 5.8 together with (7.15) and (7.16). That is, for every $\lambda \in \mathbb{R}$,

$$\mathrm{Ent}_P(\mathrm{e}^{\lambda Z}) \le \lambda^2 \sum_{i=1}^{n} \int \bigg(\iint_{\{\lambda Z_i(x_i) \ge \lambda Z_i(y_i)\}} \big[Z_i(x_i) - Z_i(y_i) \big]^2 \mathrm{e}^{\lambda Z_i(x_i)} d\mu_i(x_i) d\mu_i(y_i) \bigg) dP(x).$$

Fix i, $1 \le i \le n$. When $\lambda > 0$,

$$\begin{aligned} &\iint_{\{\lambda Z_i(x_i) \ge \lambda Z_i(y_i)\}} \big[Z_i(x_i) - Z_i(y_i) \big]^2 \mathrm{e}^{\lambda Z_i(x_i)} d\mu_i(x_i) d\mu_i(y_i) \\ &\qquad \le \iint \sum_{k=1}^{N} \tau_k(x) (x_i^k - y_i^k)^2 \mathrm{e}^{\lambda Z_i(x_i)} d\mu_i(x_i) d\mu_i(y_i), \end{aligned}$$

whereas when $\lambda < 0$,

$$\begin{aligned} &\iint_{\{\lambda Z_i(x_i) \ge \lambda Z_i(y_i)\}} \big[Z_i(x_i) - Z_i(y_i) \big]^2 \mathrm{e}^{\lambda Z_i(x_i)} d\mu_i(x_i) d\mu_i(y_i) \\ &\qquad \le \iint \sum_{k=1}^{N} \tau_k(y) (x_i^k - y_i^k)^2 \mathrm{e}^{\lambda Z_i(y_i) + 2|\lambda|} d\mu_i(x_i) d\mu_i(y_i) \end{aligned}$$

where we used (7.15) and (7.16). It follows that, for every $\lambda \in \mathbb{R}$,

$$\mathrm{Ent}_P(\mathrm{e}^{\lambda Z}) \leq \lambda^2 \mathrm{e}^{2|\lambda|} \int\int \max_{1 \leq k \leq N} \sum_{i=1}^{n} (x_i^k - y_i^k)^2 \mathrm{e}^{\lambda Z(x)} dP(x) dP(y). \tag{7.17}$$

The extra factor $\mathrm{e}^{2|\lambda|}$ with respect to analogous formulas might look annoying. However, we only use (7.17) with λ small.

Let us not recast inequality (7.17) with the original notations. Using that $(u+v)^2 \leq 2u^2 + 2v^2$ and independence, inequality (7.17) amounts to the inequality in $\lambda \in \mathbb{R}$,

$$\mathrm{Ent}_P(\mathrm{e}^{\lambda Z}) \leq 2\lambda^2 \mathrm{e}^{2|\lambda|} \Big(\mathbb{E}(\Sigma^2)\mathbb{E}(\mathrm{e}^{\lambda Z}) + \mathbb{E}(\Sigma^2 \mathrm{e}^{\lambda Z}) \Big).$$

Take a $\lambda_0 > 0$ to be specified below. In particular, for every $|\lambda| \leq \lambda_0$,

$$\mathrm{Ent}_P(\mathrm{e}^{\lambda Z}) \leq c_0 \lambda^2 \Big(\mathbb{E}(\Sigma^2)\mathbb{E}(\mathrm{e}^{\lambda Z}) + \mathbb{E}(\Sigma^2 \mathrm{e}^{\lambda Z}) \Big) \tag{7.18}$$

with $c_0 = 2\,\mathrm{e}^{2\lambda_0}$. As in Chapter 5 and Theorem 7.5 above, one has to integrate this differential inequality. We do not try here a sharp integration as in Theorem 7.5, but what follows is sufficient for the purpose of the theorem (for a more careful analysis, see [Mas1]).

We work with $\widetilde{Z} = Z - \mathbb{E}(Z)$. In terms of the Laplace transform $\widetilde{\Lambda}$ of $\widetilde{Z}$, (7.18) indicates that for every λ such that $|\lambda| \leq \lambda_0$,

$$\lambda \widetilde{\Lambda}'(\lambda) - \widetilde{\Lambda}(\lambda) \log \widetilde{\Lambda}(\lambda) \leq c_0 \lambda^2 \Big(\mathbb{E}(\Sigma^2)\widetilde{\Lambda}(\lambda) + \mathbb{E}(\Sigma^2 \mathrm{e}^{\lambda \widetilde{Z}}) \Big).$$

We first bound the term $\mathbb{E}(\Sigma^2 \mathrm{e}^{\lambda \widetilde{Z}})$. We have

$$\begin{aligned} \mathbb{E}(\Sigma^2 \mathrm{e}^{\lambda \widetilde{Z}}) &= (\mathrm{e}-1)\mathbb{E}(\Sigma^2)\mathbb{E}(\mathrm{e}^{\lambda \widetilde{Z}}) + \mathbb{E}\Big(\big[\Sigma^2 - (\mathrm{e}-1)\mathbb{E}(\Sigma^2)\big] \mathrm{e}^{\lambda \widetilde{Z}} \Big) \\ &\leq (\mathrm{e}-1)\mathbb{E}(\Sigma^2)\mathbb{E}(\mathrm{e}^{\lambda \widetilde{Z}}) + \lambda \mathbb{E}(\widetilde{Z} \mathrm{e}^{\lambda \widetilde{Z}}) - \mathbb{E}(\mathrm{e}^{\lambda \widetilde{Z}}) + \mathbb{E}\big(\mathrm{e}^{\Sigma^2 - (\mathrm{e}-1)\mathbb{E}(\Sigma^2)}\big) \end{aligned}$$

where we used Young's inequality $uv \leq u \log u - u + \mathrm{e}^v$, $u \geq 0$, $v \in \mathbb{R}$, with $u = \mathrm{e}^{\lambda \widetilde{Z}}$ and $v = \Sigma^2 - (\mathrm{e}-1)\mathbb{E}(\Sigma^2)$. Now, since $0 \leq f^2 \leq 1$, Σ^2 enters the setting of Theorem 7.5 so that $\mathbb{E}(\mathrm{e}^{\Sigma^2 - (\mathrm{e}-1)\mathbb{E}(\Sigma^2)}) \leq 1$. Since $\mathbb{E}(\mathrm{e}^{\lambda \widetilde{Z}}) \geq 1$ by Jensen's inequality, it follows that

$$\mathbb{E}(\Sigma^2 \mathrm{e}^{\lambda \widetilde{Z}}) \leq (\mathrm{e}-1)\mathbb{E}(\Sigma^2)\widetilde{\Lambda}(\lambda) + \lambda \widetilde{\Lambda}'(\lambda)$$

for every λ.

Summarizing the preceding estimates, for $|\lambda| \leq \lambda_0$,

$$\lambda \widetilde{\Lambda}'(\lambda) - \widetilde{\Lambda}(\lambda) \log \widetilde{\Lambda}(\lambda) \leq c_0 \lambda^2 \Big(\mathrm{e}\,\mathbb{E}(\Sigma^2)\widetilde{\Lambda}(\lambda) + \lambda \widetilde{\Lambda}'(\lambda) \Big).$$

We now integrate this differential inequality in the standard way. Set $H(\lambda) = \frac{1}{\lambda} \log \widetilde{\Lambda}(\lambda)$, $H(0) = 0$, $\lambda \in \mathbb{R}$. Hence,

$$H'(\lambda) \leq c_0 \bigg(\mathrm{e}\,\mathbb{E}(\Sigma^2) + \lambda \frac{\widetilde{\Lambda}'(\lambda)}{\widetilde{\Lambda}(\lambda)} \bigg).$$

We may integrate the preceding to get (recall $H(0) = 0$), for $|\lambda| \leq \lambda_0$,

$$\log \widetilde{\Lambda}(\lambda) \leq c_0\Big(\mathrm{e}\,\mathbb{E}(\Sigma^2)\lambda^2 + \lambda^2 \log \widetilde{\Lambda}(\lambda)\Big).$$

Therefore, provided that $c_0\lambda_0^2 < 1$, for $|\lambda| \leq \lambda_0$,

$$\widetilde{\Lambda}(\lambda) \leq \mathrm{e}^{\kappa_0 \mathbb{E}(\Sigma^2)\lambda^2}$$

where $\kappa_0 = \mathrm{e}\, c_0 \mathrm{e}^{2\lambda_0}(1 - c_0\lambda_0^2)^{-1}$. For $0 \leq \lambda \leq \lambda_0$ and $r \geq 0$,

$$\mathbb{P}\Big(\big\{Z - \mathbb{E}(Z) \geq r\big\}\Big) \leq \mathrm{e}^{-\lambda r + \kappa_0 \mathbb{E}(\Sigma^2)\lambda^2}.$$

Then choose $\lambda = r/2\kappa_0\mathbb{E}(\Sigma^2)$ for $r \leq 2\kappa_0\lambda_0\mathbb{E}(\Sigma^2)$ and $\lambda = \lambda_0$ for $r \geq 2\kappa_0\lambda_0\mathbb{E}(\Sigma^2)$ to get

$$\mathbb{P}\Big(\big\{Z - \mathbb{E}(Z) \geq r\big\}\Big) \leq \exp\left(-\min\left(\frac{\lambda_0 r}{2}, \frac{r^2}{4\kappa_0\mathbb{E}(\Sigma^2)}\right)\right)$$

for every $r \geq 0$. For some appropriate choice of λ_0 (e.g. $\lambda_0 = \frac{1}{5}$), and together with the same argument for $\mathbb{E}(Z) - Z$, we may state the following.

Proposition 7.7. *If $|f| \leq C$ for every f in $\mathcal{F}$, then, for all $r \geq 0$,*

$$\mathbb{P}\Big(\big\{|Z - \mathbb{E}(Z)| \geq r\big\}\Big) \leq 2\exp\left(-\frac{1}{10}\min\left(\frac{r}{C}, \frac{r^2}{3\mathbb{E}(\Sigma^2)}\right)\right).$$

We now complete the proof of Theorem 7.6 with the Poissonian bound. We use a truncation argument. For every $r \geq 0$,

$$\mathbb{P}\Big(\big\{|Z - \mathbb{E}(Z)| \geq 4r\big\}\Big) \leq \mathbb{P}\Big(\big\{|Z_\rho - \mathbb{E}(Z_\rho)| \geq r\big\}\Big) + \mathbb{P}\big(\big\{W_\rho + \mathbb{E}(W_\rho) \geq 3r\big\}\big)$$

where

$$Z_\rho = \sup_{f \in \mathcal{F}_\rho} \sum_{i=1}^n f(Y_i)$$

with $\mathcal{F}_\rho = \{f\mathbf{1}_{\{|f| \leq \rho\}}; f \in \mathcal{F}\}$, $\rho > 0$ to be determined, and

$$W_\rho = \sup_{f \in \mathcal{F}} \sum_{i=1}^n \big|f(Y_i)\big|\mathbf{1}_{\{|f(Y_i)| > \rho\}}.$$

We use Proposition 7.7 for Z_ρ to get, for every $r \geq 0$,

$$\mathbb{P}\Big(\big\{|Z_\rho - \mathbb{E}(Z_\rho)| \geq r\big\}\Big) \leq 2\exp\left(-\frac{1}{10}\min\left(\frac{r}{\rho}, \frac{r^2}{3\mathbb{E}(\Sigma^2)}\right)\right). \tag{7.19}$$

On the other hand, we apply Theorem 7.5, more precisely (7.11) to W_ρ, to get

$$\begin{aligned} \mathbb{P}\big(\big\{W_\rho + \mathbb{E}(W_\rho) \geq 3r\big\}\big) &\leq \mathbb{P}\big(\big\{W_\rho \geq \mathbb{E}(W_\rho) + r\big\}\big) \\ &\leq \exp\left(-\frac{r}{2}\log\left(1 + \frac{r}{\mathbb{E}(W_\rho)}\right)\right) \end{aligned} \tag{7.20}$$

provided that $r \geq \mathbb{E}(W_\rho)$. Choose now

$$\rho = \rho(r) = \min\left(1, \sqrt{\frac{\mathbb{E}(\Sigma^2)}{r}}\right).$$

Then

$$r \geq \mathbb{E}(W_\rho) \quad \text{and} \quad \sqrt{r\,\mathbb{E}(\Sigma^2)} \geq \mathbb{E}(W_\rho).$$

Indeed, either $\rho = 1$ and $W_\rho = 0$ or $\rho \leq 1$, i.e. $r \geq \mathbb{E}(\Sigma^2)$, so that, since $W_\rho \leq \Sigma^2/\rho$,

$$r \geq \sqrt{r\,\mathbb{E}(\Sigma^2)} = \frac{\mathbb{E}(\Sigma^2)}{\rho} \geq \mathbb{E}(W_\rho).$$

Since for every $u \geq 0$,

$$\min\left(\sqrt{u}, \frac{u}{3}\right) \geq \frac{1}{12}\log(1+4u),$$

we have by the choice of ρ

$$\min\left(\frac{r}{\rho}, \frac{r^2}{3\mathbb{E}(\Sigma^2)}\right) \geq \frac{r}{12}\log\left(1 + \frac{4r}{\mathbb{E}(\Sigma^2)}\right)$$

while

$$\begin{aligned}\log\left(1 + \frac{r}{\mathbb{E}(W_\rho)}\right) &\geq \log\left(1 + \sqrt{\frac{r}{\mathbb{E}(\Sigma^2)}}\right)\\ &\geq \frac{1}{4}\log\left(1 + \frac{4r}{\mathbb{E}(\Sigma^2)}\right).\end{aligned}$$

Therefore, as a consequence of (7.19) and (7.20),

$$\mathbb{P}\Big(\big\{|Z - \mathbb{E}(Z)| \geq 4r\big\}\Big) \leq 3\exp\left(-\frac{r}{120}\log\left(1 + \frac{4r}{\mathbb{E}(\Sigma^2)}\right)\right).$$

Changing r into $r/4$ completes the proof of Theorem 7.6. □

Together with (7.10), Theorem 7.6 gives rise to the following more useful version. Recall that when we deal with the supremum $Z = \sup_{f\in\mathcal{F}} \sum_{i=1}^n f(Y_i)$ without absolute values, we set

$$\overline{Z} = \sup_{f\in\mathcal{F}}\left|\sum_{i=1}^n f(Y_i)\right|.$$

Corollary 7.8. *If $|f| \leq C$ for every f in $\mathcal{F}$, and if $\mathbb{E}(f(Y_i)) = 0$ for every $f \in \mathcal{F}$ and $i = 1, \ldots, n$, then, for all $r \geq 0$,*

$$\mathbb{P}\Big(\big\{|Z - \mathbb{E}(Z)| \geq r\big\}\Big) \leq 3\exp\left(-\frac{r}{KC}\log\left(1 + \frac{Cr}{\sigma^2 + C\mathbb{E}(\overline{Z})}\right)\right)$$

where $\sigma^2 = \sup_{f\in\mathcal{F}} \sum_{i=1}^n \mathbb{E}f^2(Y_i)$ and $K > 0$ is a numerical constant.

If one is not concerned with Poissonian behavior, the Bernstein inequality of Proposition 7.7 may also be stated in the following more tractable way (see [Mas1], [Mas2]). Assume for simplicity that $Z = \overline{Z}$.

Corollary 7.9. *If $|f| \leq C$ for every f in $\mathcal{F}$, and if $\mathbb{E}(f(Y_i)) = 0$ for every $f \in \mathcal{F}$ and $i = 1, \dots, n$, then, for all $r \geq 0$ and $\varepsilon > 0$,*

$$\mathbb{P}\Big(\big\{Z \geq (1+\varepsilon)\mathbb{E}(Z) + \sigma\sqrt{Kr} + K(1+\varepsilon^{-1})Cr\big\}\Big) \leq \mathrm{e}^{-r}$$

and

$$\mathbb{P}\Big(\big\{Z \leq (1-\varepsilon)\mathbb{E}(Z) - \sigma\sqrt{Kr} - K(1+\varepsilon^{-1})Cr\big\}\Big) \leq \mathrm{e}^{-r}$$

where $K > 0$ is numerical.

As announced, sharp values of the numerical constant K may be obtained by a more careful analysis. In particular, the optimal value of the constant in front of the variance term σ for the first inequality of Corollary 7.9 is achieved in [Ri3] by the entropic method in the identically distributed case.

Notes and Remarks

Bounds on norms of sums of independent random vectors or empirical processes motivated the early investigation by M. Talagrand of concentration inequalities in product spaces. Applications of Gaussian isoperimetry to sharp bounds on Gaussian random vectors and processes, initiated in [Bor2], were fully developed in the seventies after the early integrability theorems by H. J. Landau and L. A. Shepp [L-S] and X. Fernique [Fe] (cf. [Le-T], [Le3], [Li], [Bog], [Fe], etc. for a description of the historical developments of the Gaussian theory leading to Theorem 7.1 and Corollary 7.2). As alluded to in Section 7.1, the deviation and concentration from or around the mean or median of supremum of Gaussian processes do not lead in general to estimates on the mean or median themselves. These have to be handled by other means such as entropy or majorizing measures, which may also be used to yield directly, for more explicit Gaussian processes, sharp deviation inequalities (cf. [Le-T], [Li], [Tal5], [Tal11], etc.). The integrability result (7.6) is due to S. Kwapień [Kw]. The Khintchine-Kahane inequalities (7.7) go back to [Ka]. Sharp constants are discussed in [K-L-O].

The early Gaussian study together with some crucial open problems on limit theorems for sums of independent random vectors (in particular the law of the iterated logarithm) prompted M. Talagrand to investigate from an abstract measure theoretic point of view concentration properties in product spaces. This led him, after a first contribution on the discrete cube [Tal1] (that covers Theorem 7.3 in this case), to the breakthrough [Tal2] on which most of the further developments are based. These results allowed the monograph [Le-T] that solved with these tools most of the open questions on strong limit theorems for Banach space valued random variables with the approach presented here in Section 7.2 (see also [Le-Z]). Theorem 7.4 is taken from [Le-T] and [Tal5]. For statistical applications, cf. [V-W]. The isoperimetric inequality of [Tal2], and its application to a bound on sums of independent random vectors, then received a number of improvements and simplified proofs ([Tal4], [Tal7]) which finally led to the memoir [Tal7] that crowns with definitive results and arguments this deep investigation of isoperimetric and concentration inequalities for product measures.

To answer the questions by L. Birgé and P. Massart on concentration inequalities around the mean or median of supremum of empirical processes motivated by applications to selection of models in statistics [B-M1], [B-M2], [B-B-M] (cf. [Mas2]), M. Talagrand undertook in [Tal9] a further technical refinement of his methods to reach the rather definitive Theorem 7.6. In [Le4], a simplified approach to this result is presented with the tool of logarithmic Sobolev inequalities and the Herbst argument (entropic method) exposed in Section 5.2. The argument of [Le4] has been carefully examined in [Mas1] in order to reach numerical constants of statistical use. The optimal Theorem 7.5 is in particular taken from [Mas1] (see also [Mas2]). Recently, E. Rio [Ri2] further improves the entropic method, by clever integration of the differential inequality, to binomial (in the spirit of Hoeffding's inequalities [Hoe]) rather than Poissonian bounds of empirical processes based on classes of indicator functions (see also [A-V] for large deviation results). A parallel investigation of exponential integrability of sums of independent random vectors relying on hypercontractive methods is undertaken in [Kw-S] (see [K-W] for a presentation in the book form).

8. SELECTED APPLICATIONS

In this chapter, we further illustrate with a few applications the concentration of measure phenomenon. While these often lie at some mild level, the power of the concentration ideas is their wide range of potential usefulness. The first section is devoted to concentration for harmonic measures on the sphere proved by G. Schechtman and M. Schmuckenschläger [S-S2]. In Section 8.2, we present recent work of C. McDiarmid [MD3] on concentration for independent permutations extending prior contribution in [Tal7]. We next describe applications due to M. Talagrand [Tal7], mainly without proofs, of the convex hull approximation of Section 4.2 to various problems in discrete algorithmic mathematics, such as tail estimates on the longest increasing subsequence of a sample of independent uniformly distributed random variables on the unit interval, the traveling salesman problem, the minimum spanning tree, first passage times in percolation and the assignment problem. In Section 8.4, we present recent developments by M. Talagrand [Tal12] on tails of the free energy function in the Sherrington-Kirkpatrick spin glass model. Finally we briefly mention in the last part some concentration results for the spectral measure of large random matrices.

8.1 Concentration of harmonic measures

Denote by σ^n normalized Lebesgue measure on the unit sphere $\mathbb{S}^n$ in $\mathbb{R}^{n+1}$. For a point x in the open unit ball $\mathcal{B}^{n+1}$ of $\mathbb{R}^{n+1}$, denote by σ_x^n the probability measure on $\mathbb{S}^n$ given by

$$d\sigma_x^n(y) = \frac{1 - |x|^2}{|y - x|^{n+1}} \, d\sigma^n(y) \tag{8.1}$$

where we recall that $|\cdot|$ is the standard Euclidean norm. We denote below by $x \cdot y$ the scalar product of $x, y \in \mathbb{R}^{n+1}$. If f is integrable on $\mathbb{S}^n$, $\int f d\sigma_x^n$ is harmonic in $\mathcal{B}^{n+1}$ with radial limits equal σ^n-almost everywhere to f.

The measures σ_x^n are the so-called *harmonic measures* on the sphere $\mathbb{S}^n$. They have a neat probabilistic interpretation in terms of exit times of Brownian motion. Indeed, if $\mathbb{P}_x$ is the probability distribution of a standard Brownian motion $(W_t)_{t \geq 0}$ in $\mathbb{R}^{n+1}$ starting from $x \in \mathcal{B}^{n+1}$, and if T is the first time t for which W_t hits $\mathbb{S}^n$, then the distribution of W_T under $\mathbb{P}_x$ is σ_x^n.

The following theorem has been proved in [S-S2] and extends concentration of the sphere to all harmonic measures uniformly over x.

Theorem 8.1. *For every* $|x| < 1$,

$$\alpha_{(\mathbb{S}^n, d, \sigma_x^n)}(r) \leq 4\,\mathrm{e}^{-cnr^2}, \quad r > 0,$$

where $c > 0$ *is numerical.*

Proof. It is divided in several steps.

Lemma 8.2. *For all* $|x| < 1$ *and* $0 \leq \lambda \leq \frac{(n+1)^2}{8}$,

$$\mathbb{E}_x\big(\mathrm{e}^{\lambda T}\big) \leq \mathrm{e}^{2\lambda(1-|x|^2)/(n+1)}.$$

Proof. The function $u(y) = \mathrm{e}^{-\lambda|y|^2}$, $y \in \mathbb{R}^{n+1}$, satisfies the equation $\frac{1}{2}\,\Delta u + cu = 0$ where $c = c(y) = \lambda(n+1) - 2\lambda^2|y|^2$. It thus follows from Itô's formula ([Dur], [R-Y], etc.) that

$$M_t = \exp\bigg(- \lambda|W_t|^2 + \int_0^t c(W_s)ds\bigg), \quad t \geq 0,$$

is a martingale. In particular, for every integer $k \geq 0$,

$$\mathrm{e}^{\lambda|y|^2} = \mathbb{E}_x(M_0) = \mathbb{E}_x(M_{T\wedge k}).$$

Now, for $0 \leq t \leq T$ and $0 < \lambda \leq \frac{n+1}{4}$, we have

$$2\lambda^2|W_t|^2 \leq \frac{\lambda(n+1)}{2}$$

so that

$$\mathbb{E}_x\big(\mathrm{e}^{\lambda(n+1)T/2}\mathbf{1}_{\{T\leq k\}}\big) \leq \mathrm{e}^{\lambda(1-|x|^2)}.$$

Replacing λ by $2\lambda/(n+1)$ yields the conclusion. □

For $0 < |x| \leq 1$, let

$$A(x) = \Big\{x + \big(1 - |x|^2\big)^{1/2} z; |z| = 1, z \perp x\Big\}.$$

The next geometric lemma is left to the reader (see [S-S2] for the details).

Lemma 8.3. *Let* $0 < |x| \leq 1$. *For any* $y \in \mathbb{S}^n$, *the Euclidean distance* $d(y, A(x))$ *satisfies*

$$d\big(y, A(x)\big) \leq 2\,\frac{|(y-x)\cdot\frac{x}{|x|}|}{(1-|x|^2)^{1/2}}$$

if $(y-x)\cdot x > 0$ *whereas when* $(y-x)\cdot x < 0$,

$$d\big(y, A(x)\big) \leq 2\,\frac{|(y-x)\cdot\frac{x}{|x|}|}{(1-(y\cdot\frac{x}{|x|})^2)^{1/2}}\,.$$

One shows similarly that

$$d\big(y, A(x)\big) \leq 2\left|(y-x)\cdot\frac{x}{|x|}\right|^{1/2}. \tag{8.2}$$

For $0 < |x| \leq 1$ and $-1 < \lambda < +1$, set $\mathbb{S}_\lambda^{n-1}(x) = A\big(\lambda \frac{x}{|x|}\big)$.

Proposition 8.4. *Let $0 < |x| \leq 1$ and let $F : \mathbb{S}^n \to \mathbb{R}$ be a 1-Lipschitz function which is constant on each $\mathbb{S}_\lambda^{n-1}(x)$, $-1 < \lambda < +1$, $n \geq 2$. Then for some constant $a_F \in \mathbb{R}$,*

$$\sigma_x^n\big(\{|F - a_F| \geq r\}\big) \leq 2\,\mathrm{e}^{-(n+1)r^2/32}$$

for every $r \geq 0$.

Proof. Let a_F be the constant value of F on $\mathbb{S}_{|x|}^{n-1}(x) = A(x)$. We may and do assume that $a_F = 0$. Note that by (8.2),

$$\big\{|F| \geq r\big\} \subset \big\{y; d\big(y, A(x)\big) \geq r\big\} \subset \Big\{y; \Big|(y-x)\cdot\frac{x}{|x|}\Big| \geq \frac{r^2}{2}\Big\}.$$

To evaluate the measure of the last set, we again apply Itô's formula. For $\lambda \in \mathbb{R}$, consider now $u(y) = \mathrm{e}^{\lambda(y-x)\cdot\frac{x}{|x|}}$, $y \in \mathbb{R}^{n+1}$. Then $\Delta u = \lambda^2 u$ and thus

$$M_t = \exp\bigg(\lambda(W_t - x)\cdot\frac{x}{|x|} - \frac{\lambda^2}{2}\,t\bigg), \quad t \geq 0,$$

is a bounded martingale under $\mathbb{P}_x$. In particular $\mathbb{E}_x(M_T) = M_0 = 1$ and thus

$$\begin{aligned}
&\mathbb{E}_x\big(\mathrm{e}^{\lambda(W_T - x)\cdot\frac{x}{|x|}}\big)\\
&\quad= \mathbb{E}_x\big(\mathrm{e}^{\lambda(W_T - x)\cdot\frac{x}{|x|} - \lambda^2 T}\mathrm{e}^{\lambda^2 T}\big)\\
&\quad\leq \Big(\mathbb{E}_x\big(\mathrm{e}^{2\lambda(W_T - x)\cdot\frac{x}{|x|} - \frac{(2\lambda)^2}{2}T}\big)\Big)^{1/2}\Big(\mathbb{E}_x\big(\mathrm{e}^{2\lambda^2 T}\big)\Big)^{1/2}\\
&\quad\leq \mathrm{e}^{2\lambda^2(1-|x|^2)/(n+1)}
\end{aligned}$$

for $\lambda \leq \frac{n+1}{4}$ from Lemma 8.2.

Since the law of W_T under $\mathbb{P}_x$ is σ_x^n, when $r > 2(1-|x|^2)^{1/2}$, we may choose $\lambda = \frac{n+1}{4}$ in Chebyshev's exponential inequality to get from the preceding that

$$\sigma_x^n\big(\{|F| \geq r\}\big) \leq 2\,\mathrm{e}^{-(n+1)r^2/8}.$$

When $r \leq 2(1-|x|^2)^{1/2}$ we use the bounds of Lemma 8.3 to get that

$$\begin{aligned}
\sigma_x^n\big(\{|F| \geq r\}\big) &\leq \sigma_x^n\Big(\Big\{y; (y-x)\cdot\frac{x}{|x|} \geq \frac{r}{2}\big(1-|x|^2\big)^{1/2}\Big\}\Big)\\
&\leq 2\exp\bigg(-\frac{\lambda r(1-|x|^2)^{1/2}}{2} + \frac{2\lambda^2(1-|x|^2)}{n+1}\bigg)
\end{aligned}$$

for all $0 < \lambda \leq \frac{n+1}{4}$. Choosing $8\lambda(1-|x|^2)^{1/2} = (n+1)r$, we get

$$\sigma_x^n\big(\{|F| \geq r\}\big) \leq 2\,\mathrm{e}^{-(n+1)r^2/32}.$$

The proof of the proposition is complete. □

We may now complete the proof of Theorem 8.1. Assume that $0 < |x| < 1$. Let F be 1-Lipschitz on $\mathbb{S}^n$, $n \geq 3$. For $-1 < \lambda < +1$, denote by $a(\lambda)$ the expectation of F on $\mathbb{S}^{n-1}_\lambda(x)$ with respect to σ^n_x conditioned on $\mathbb{S}^{n-1}_\lambda(x)$ (that can be viewed as normalized Haar measure σ^{n-1} on $\mathbb{S}^n \cap x^\perp$). Now,

$$F_\lambda(z) = F\Big(\lambda \frac{x}{|x|} + (1-\lambda^2)^{1/2} z\Big)$$

is Lipschitz with Lipschitz constant $(1-\lambda^2)^{1/2} \leq 1$ on $\mathbb{S}^n \cap x^\perp$. Concentration on spheres (Theorem 2.3) thus yields

$$\sigma^n_x\Big(\big\{y; \big|F(y) - a\big(y, \tfrac{x}{|x|}\big)\big| \geq r\big\}\Big) \leq 2\,\mathrm{e}^{-(n-2)r^2/8}.$$

Now, $a(y \cdot \frac{x}{|x|})$ is 1-Lipschitz on $\mathbb{S}^n$ and by Proposition 8.4, for some $a \in \mathbb{R}$ and all $r \geq 0$,

$$\sigma^n_x\Big(\big\{y; \big|a\big(y, \tfrac{x}{|x|}\big) - a\big| \geq r\big\}\Big) \leq 2\,\mathrm{e}^{-c(n+1)r^2/32}.$$

It easily follows from the triangle inequality that

$$\sigma^n_x\big(\{|F-a| \geq r\}\big) \leq 4\,\mathrm{e}^{-c(n-2)r^2}, \quad r \geq 0,$$

for some numerical $c > 0$. By Proposition 1.8, we may then come to the result of Theorem 8.1. It is not difficult to check that the theorem also holds for $n = 1, 2$. The proof is complete. □

Harmonic measures are related to image measures of uniform measure on $\mathbb{S}^n$ under the so-called Möbius transformations. A modification of the proof of Theorem 8.1 then gives rise to an improvement of the deviation inequalities for Lipschitz functions by replacing the Lipschitz coefficient of a function F by the infimum over the Lipschitz constants of F composed with all Möbius transformations.

If x is a point in the Euclidean (open) unit ball $\mathcal{B}^{n+1}$ of $\mathbb{R}^{n+1}$, denote by P_x the orthogonal projection onto the subspace generated by x, and by Q_x the orthogonal projection onto the subspace orthogonal to x. Define then the Möbius transformation ϕ_x by

$$\phi_x(y) = \frac{x - P_x(y) - (1-|x|^2)^{1/2} Q_x(y)}{1 - (y \cdot x)}.$$

It is known (cf. [Ru]) that ϕ_x is an involutive diffeomorphism of $\mathcal{B}^{n+1}$ onto itself, that ϕ_x restricted to $\mathbb{S}^n$ is an involutive diffeomorphism of $\mathbb{S}^n$ onto itself and that for every y in the closure of $\mathcal{B}^{n+1}$,

$$1 - \big|\phi_x(y)\big|^2 = \frac{(1-|x|^2)(1-|y|^2)}{(1-(y \cdot x))^2}.$$

Proposition 8.4 may be adapted (a more direct approach is also suggested in [S-S2]), to yield the same conclusion for the image measures of $\mathbb{S}^n$ under ϕ_x. One may then derive, as for Theorem 8.1, the following sharpening of the concentration phenomenon on $\mathbb{S}^n$.

Theorem 8.5. *Let F be a real continuous function on $\mathbb{S}^n$, $n \geq 2$, such that* $\inf \|F \circ \phi_x\|_{\mathrm{Lip}} \leq 1$ *where the infimum is running over all x's in the unit ball of $\mathbb{R}^{n+1}$. Then, for every $r \geq 0$,*

$$\sigma^n\big(\{|F - \textstyle\int F d\sigma| \geq r\}\big) \leq 4\,\mathrm{e}^{-cnr^2}$$

where $c > 0$ is numerical.

8.2 Concentration for independent permutations

This section is taken from a recent work by C. McDiarmid [MD3], which improves upon some early result in [Tal7]. It aims to present a uniform version of the concentration property on the symmetric group Π^n of Corollary 4.3 in the same way Theorem 4.6 extends concentration for the Hamming metric.

We start with some notation. Recall the symmetric group Π^n over $\{1, \ldots, n\}$. Let $B_1 \cup B_2 \cup \cdots \cup B_\ell$ be a partition of $\{1, \ldots, n\}$ and set $G = \prod_{k=1}^{\ell} \Pi(B_k)$ where $\Pi(B_k)$ is the symmetric group over B_k. We denote by P the product measure on G of uniform probability measures on each $\Pi(B_k)$. Thus P is uniform over G. We may consider the convex functional $\mathcal{D}_A^c(\sigma)$ of Section 4.2 on Π^n with thus $\sigma \in \Pi^n$, and A non-empty subset of Π^n. In particular, as we have seen there,

$$\mathcal{D}_A^c(\sigma) = \inf_{y \in V_A(\sigma)} |y|$$

where for a permutation $\sigma = (\sigma(1), \ldots, \sigma(n))$ in Π^n and A a non-empty subset of Π^n, $V_A(\sigma)$ denotes the convex hull of the vectors $(\mathbf{1}_{\{\sigma(i) \neq \pi(i)\}})_{1 \leq i \leq n}$, $\pi \in A$, in $[0,1]^n$. We also recall that $|y|$ denotes here the Euclidean norm $\big(\sum_{i=1}^n y_i^2\big)^{1/2}$ of $y = (y_1, \ldots, y_n) \in \mathbb{R}^n$.

Theorem 8.6. *Let G be a product group of permutations on the set $\{1, \ldots, n\}$ and let P be a uniform measure on G. Then, for every non-empty subset $A \subset G$,*

$$\int \mathrm{e}^{(\mathcal{D}_A^c)^2/16} dP \leq \frac{1}{P(A)}\,.$$

Before turning to the proof of the theorem, we present some useful lemmas.

The first lemma shows that $V_A(\sigma)$ is stable under composition and inverse. If $\tau \in \Pi^n$, $V_A(\sigma)\tau$ denotes the collection of all $(v_{\tau(i)})_{1 \leq i \leq n}$, $v \in V_A(\sigma)$.

Lemma 8.7. *For $A \subset \Pi^n$ and $\sigma, \pi \in \Pi^n$,*

$$V_{A\pi}(\sigma\pi) = V_A(\sigma)\pi \quad \textit{and} \quad V_{A^{-1}}(\sigma^{-1}) = V_A(\sigma)\sigma^{-1}.$$

Proof. Simply note that if $s = (s_i)_{1 \leq i \leq n}$ is defined by

$$s_i = \mathbf{1}_{\{\sigma(i) \neq \tau(i)\}}, \quad i = 1, \ldots, n,$$

for some $\tau \in A$, then $s\pi = (s_{\pi(i)})_{1 \leq i \leq n}$ is defined by

$$s_{\pi(i)} = \mathbf{1}_{\{\sigma\pi(i) \neq \tau\pi(i)\}}, \quad i = 1, \ldots, n,$$

where $\tau\pi \in A\pi$. The first identity follows. The second is proved similarly. □

Besides $\mathcal{D}_A^c(\sigma)$, define for $i \in \{1, \dots, n\}$,

$$\mathcal{D}_A^c(\sigma, i) = \inf_{y \in V_A(\sigma)} \left(|y|^2 + y_i^2\right)^{1/2}.$$

Observe that $\mathcal{D}_A^c(\sigma, i) = \mathcal{D}_A^c(\sigma)$ if $\pi(i) = \sigma(i)$ for each $\pi \in A$, and in particular if $\pi(i) = i$ for each $\pi \in A \cup \{\sigma\}$. As a consequence, the preceding lemma immediately yields the following.

Lemma 8.8. *For $A \subset \Pi^n$, $\sigma, \pi \in \Pi^n$, and $i \in \{1, \dots, n\}$,*

$$\mathcal{D}_A^c(\sigma) = \mathcal{D}_{A\pi}^c(\sigma\pi), \quad \mathcal{D}_A^c(\sigma, i) = \mathcal{D}_{A\pi}^c\left(\sigma\pi, \pi^{-1}(i)\right)$$

and

$$\mathcal{D}_A^c(\sigma, i) = \mathcal{D}_{A^{-1}}^c\left(\sigma^{-1}, \sigma(i)\right).$$

The next lemma is crucial in the induction proof of Theorem 8.6. It is similar to the argument used for Theorem 4.6.

Lemma 8.9. *Let $A \subset \Pi^n$, $\sigma \in \Pi^n$, $i \neq j$ in $\{1, \dots, n\}$ and let $0 \leq \theta \leq 1$. Set $A_i = \{\pi \in A; \pi(i) = \sigma(i)\}$ and suppose that A_i and A_j are non-empty. Then*

$$\mathcal{D}_A^c(\sigma, i)^2 \leq 4(1-\theta)^2 + \theta\mathcal{D}_{A_i}^c(\sigma, j)^2 + (1-\theta)\mathcal{D}_{A_j\tau_{ij}}^c(\sigma)^2$$

where τ_{ij} is the permutation that exchanges i and j.

Proof. Given $i \neq j$ in $\{1, \dots, n\}$, set further

$$\mathcal{D}_A^c(\sigma, i, j) = \inf_{y \in V_A(\sigma)} \left(|y|^2 - y_i^2 - y_j^2\right)^{1/2}.$$

Let $\xi \in V_A(\sigma)$ satisfy $\xi_i = 0$ and $\mathcal{D}_{A_i}^c(\sigma, j)^2 = |\xi|^2 + \xi_j^2$ and let $\zeta \in V_A(\sigma)$ satisfy $\mathcal{D}_A^c(\sigma, i, j)^2 = |\zeta|^2 - \zeta_i^2 - \zeta_j^2$. Since $V_A(\sigma)$ is convex, the point $v = \theta\xi + (1-\theta)\zeta$ is in $V_A(\sigma)$. Thus

$$\mathcal{D}_A^c(\sigma, i)^2 \leq |v|^2 + v_i^2.$$

Furthermore, by the convexity property of the square function, for every $1 \leq k \leq n$,

$$v_k^2 \leq \theta\xi_k^2 + (1-\theta)\zeta_k^2.$$

Also,

$$2v_i^2 = 2(1-\theta)^2\zeta_i^2 \leq 2(1-\theta)^2$$

and

$$v_j^2 \leq 2\theta^2\xi_j^2 + 2(1-\theta)^2\zeta_j^2 \leq 2\theta\xi_j^2 + 2(1-\theta)^2.$$

Hence,

$$\begin{aligned}
\mathcal{D}_A^c(\sigma, i)^2 &\leq 4(1-\theta)^2 + 2\theta\xi_j^2 + \sum_{k \neq i,j} \left(\theta\xi_k^2 + (1-\theta)\zeta_k^2\right) \\
&= 4(1-\theta)^2 + \theta\left(|\xi|^2 + \xi_j^2\right) + (1-\theta)\left(|\zeta|^2 - \zeta_i^2 - \zeta_j^2\right) \\
&= 4(1-\theta)^2 + \theta\mathcal{D}_{A_i}^c(\sigma, j)^2 + (1-\theta)\mathcal{D}_A^c(\sigma, i, j)^2.
\end{aligned}$$

Finally note that

$$\mathcal{D}_A^c(\sigma, i, j) \le \mathcal{D}_{A\tau_{ij}}^c(\sigma) \le \mathcal{D}_{A_i\tau_{ij}}^c(\sigma)$$

from which the conclusion follows. The lemma is established. □

We turn to the proof of Theorem 8.6. We use induction on the order of G to prove a slightly stronger result, namely that for each i in $\{1, \dots, n\}$,

$$\int e^{(\mathcal{D}_A^c(\cdot,i))^2/16} dP \le \frac{1}{P(A)} \,. \tag{8.3}$$

The inequality is obvious if G is reduced to the identity permutation since then $A = G$. It suffices also to assume that i is such that $\pi(i) \neq i$ for some $\pi \in G$ since for each $j \in \{1, \dots, n\}$ for which this is not the case,

$$\mathcal{D}_A^c(\sigma, j) = \mathcal{D}_A^c(\sigma) \le \mathcal{D}_A^c(\sigma, i)$$

for every $\sigma \in G$. We may thus assume, to establish (8.3), that $\pi(i) \neq i$ for some $\pi \in G$ and that $i = 1$.

Assume now that G is non-trivial and suppose (8.3) holds for any product group which is a proper subgroup of G. Then let $H = \{\sigma \in G; \sigma(1) = 1\}$ be the stabilizer of the element 1. We may assume without loss of generality that the orbit $\{\sigma(1); \sigma \in G\}$ is $\{1, \dots, m\}$ for some $m \ge 2$. Then H is a product group of permutations of Π^n where the block $\{1, \dots, m\}$ has been split into the two blocks $\{1\}$ and $\{2, \dots, m\}$. Let $H_1 = H$ and for $i = 1, \dots, m$, let H_i denote the coset $H\tau_{1i} = \{\sigma\tau_{1i}; \sigma \in H\}$. Thus G is partitioned into the m cosets H_i where H_i consists of the permutations σ with $\sigma(i) = 1$ and each H_i has size $m^{-1}\mathrm{Card}(G)$.

To ease the notation, we set, for a non-empty subset $A \subset \Pi^n$,

$$\phi_A(\sigma) = \frac{1}{16}\left(\mathcal{D}_A^c(\sigma)\right)^2 \quad \text{and} \quad \phi_A(\sigma, i) = \frac{1}{16}\left(\mathcal{D}_A^c(\sigma, i)\right)^2.$$

Now let $A \subset G$, and for $i = 1, \dots, m$, let $A_i = A \cap H_i$. Choose j such that $P(A_j)/P(H)$ is maximum, and keep j fixed. For every $i \in \{1, \dots, m\}$, denote by P_i a uniform probability measure on H_i. The main part of the proof is to establish that, for every i,

$$\int e^{\phi_A(\cdot,i)} dP_i \le \frac{P(H)}{P(A_j)}\left(2 - \frac{P(A_i)}{P(A_j)}\right). \tag{8.4}$$

Let us agree that τ_{11} is the identity element. By Lemma 8.8,

$$\int e^{\phi_{A_i}(\cdot,j)} dP_i = \int e^{\phi_{A_i\tau_{1i}}(\sigma\tau_{1i},\tau_{1i}(j))} dP_i(\sigma)$$

so that

$$\int e^{\phi_{A_i}(\cdot,j)} dP_i \le \frac{P(H)}{P(A_i)} \tag{8.5}$$

by the induction hypothesis since $A_i\tau_{1i} \subset H$ and $P(A_i\tau_{1i}) = P(A_i)$. By Lemma 8.8 again,

$$\int e^{\phi_{A_j\tau_{ij}}} dP_i = \int e^{\phi_{A_j\tau_{ij}\tau_{1i}}(\sigma\tau_{1i})} dP_i(\sigma)$$

and hence

$$\int e^{\phi_{A_j\tau_{ij}}}\,dP_i \le \frac{P(H)}{P(A_j)} \tag{8.6}$$

by the induction hypothesis since $A_j\tau_{ij}\tau_{1i} \subset H$ and $P(A_j\tau_{ij}\tau_{1i}) = P(A_j)$.

Suppose first that $i \neq j$. By Lemma 8.9, for every $0 \le \theta \le 1$, as in the proof of Theorem 4.6,

$$\begin{aligned}\int e^{\phi_A(\sigma,i))}dP_i(\sigma) &\le e^{(1-\theta)^2/4}\int e^{\theta\phi_{A_i}(\sigma,j)+(1-\theta)\phi_{A_j\tau_{ij}}(\sigma)}dP_i(\sigma)\\ &\le e^{(1-\theta)^2/4}\Big(\int e^{\phi_{A_i}(\sigma,j)}dP_i(\sigma)\Big)^{\theta}\Big(\int e^{\phi_{A_j\tau_{ij}}}dP_i\Big)^{1-\theta}\\ &\le e^{(1-\theta)^2/4}\Big(\frac{P(H)}{P(A_i)}\Big)^{\theta}\Big(\frac{P(H)}{P(A_j)}\Big)^{1-\theta}\end{aligned}$$

where we used Hölder's inequality and (8.5), (8.6). Minimizing over θ using (4.7) yields the claim (8.4) in this case. The case $i = j$ immediately follows from (8.5).

We may now conclude the proof of the theorem on the basis of (8.4). Write

$$\int e^{\phi_A(\sigma,\sigma^{-1}(1))}dP(\sigma) = \frac{1}{m}\sum_{i=1}^{m}\int e^{\phi_A(\sigma,i)}dP_i(\sigma).$$

Hence, by (8.4),

$$\begin{aligned}\int e^{\phi_A(\sigma,\sigma^{-1}(1))}dP(\sigma) &\le \frac{1}{m}\sum_{i=1}^{m}\frac{P(H)}{P(A_j)}\Big(2-\frac{P(A_i)}{P(A_j)}\Big)\\ &= \frac{P(H)}{P(A_j)}\Big(2-\frac{P(A)P(H)}{P(A_j)}\Big)\\ &\le \frac{1}{P(A)}\end{aligned}$$

since $u(2-u) \le 1$ for any real number u. Replace now A by A^{-1}. By Lemma 8.8, $\mathcal{D}_A^c(\sigma,1)$ and $\mathcal{D}_{A^{-1}}^c(\sigma^{-1},\sigma(1))$ have the same distribution under P, from which the desired claim follows since $P(A^{-1}) = P(A)$. The proof of Theorem 8.6 is complete. □

Theorem 8.6 was motivated by randomized methods for graph coloring, and in particular by bounds on the chromatic number of a graph in terms of the clique number and the maximum degree (cf. [MD3], [M-R]). To this task, it may be coupled with the product space statement (Theorem 4.6) to yield concentration inequalities for certain classes of functionals of interest in graph theory. Let $X = X_1 \times \cdots \times X_n$ be a product space equipped with a product probability measure $Q = \mu_1 \otimes \cdots \otimes \mu_n$. Recall the product group $G = \prod_{k=1}^{\ell}\Pi(B_k)$ where $B_1 \cup \cdots \cup B_\ell$ is a partition of $\{1,\dots,n\}$ and $\Pi(B_k)$ is the symmetric group over B_k. As before, let P be the product measure on G of uniform probability measures on each $\Pi(B_k)$. We may then consider a function $F = F(x,\sigma)$ on the product space $X \times G$. Assume first that F is 1-Lipschitz with respect to the Hamming metric d on X and G in the sense that for all $(x,\sigma), (y,\pi)$ in (X,G),

$$\big|F(x,\sigma) - F(y,\pi)\big| \le 2d(x,y) + d(\sigma,\pi). \tag{8.7}$$

Assume moreover that F is t-determined for some $t > 0$, in the sense that whenever $F(x, \sigma) = s$, then any $(y, \pi) \in X \times G$ that agrees with (x, σ) on at least ts coordinates is such that $F(y, \pi) \geq s$. The following consequence of both Theorems 8.6 and 4.6 (cf. [MD3]) is typically used with t small, say 1 or 2, to show that such an F is strongly concentrated around a median or mean.

Theorem 8.10. *Let F on $X \times G$ be t-determined and 1-Lipschitz in the sense of (8.7), and let m be a median of F for $Q \otimes P$. Then, for each $r \geq 0$,*

$$Q \otimes P\big(\{F \geq m + r\}\big) \leq 2\,\mathrm{e}^{-r^2/16t(m+r)}$$

and

$$Q \otimes P\big(\{F \leq m - r\}\big) \leq 2\,\mathrm{e}^{-r^2/16t}.$$

In particular,

$$Q \otimes P\big(\{|F - m| \geq r\}\big) \leq 4\,\mathrm{e}^{-r^2/32t}$$

for $0 \leq r \leq m$.

8.3 Subsequences, percolation, assignment

This section is devoted to applications of Theorem 4.6 to various questions in discrete and combinatorial probability theory. These applications are due to M. Talagrand and are taken, mostly without proofs, from the memoir [Tal7] (see also [MD2], [Ste]). While the results of this section all produce strong concentration around some mean value, they do not control the respective size of the functionals under study.

We first examine subsequences. Consider points $x_1, \ldots, x_n$ in $[0,1]$. Denote by $L_n(x_1, \ldots, x_n)$ the length of the longest increasing subsequence of $x_1, \ldots, x_n$, that is, the largest integer p such that we can find $i_1 < \cdots < i_p$ for which $x_{i_1} \leq \cdots \leq x_{i_p}$. It is easy to see that when $U_1, \ldots, U_n$ are independent uniformly distributed over $[0,1]$, the random variable $L_n(U_1, \ldots, U_n)$ is distributed like the length of the longest increasing subsequence of a random permutation of $\{1, \ldots, n\}$. Concentration properties for $L_n(U_1, \ldots, U_n)$ may be obtained as a simple consequence of the convex hull approximation studied in Section 4.2.

For $x = (x_1, \ldots, x_n) \in [0,1]^n$, denote more simply $L_n(x) = L_n(x_1, \ldots, x_n)$. Recall the convex hull approximation functional $\mathcal{D}_A^c$ from Section 4.2.

Lemma 8.11. *For all $s \geq 0$ and $x \in [0,1]^n$,*

$$s \geq L_n(x) - \mathcal{D}_A^c(x)\sqrt{L_n(x)}$$

where $A = \{L_n \leq s\}$. In particular,

$$\mathcal{D}_A^c(x) \geq \frac{u}{\sqrt{s+u}}$$

whenever $L_n(x) \geq s + u$.

Proof. By definition, we can find a subset I of $\{1, \ldots, n\}$ of cardinality $L_n(x)$ such that if $i, j \in I$, $i < j$, then $x_i < x_j$. It follows from the definition of $\mathcal{D}_A^c$ that there exists $y \in A$ such that

$$\mathrm{Card}\,(J) \leq \mathcal{D}_A^c\sqrt{L_n(x)}$$

where $J = \{i \in I; y_i \neq x_i\}$. Thus $(x_i)_{i \in I \setminus J}$ is an increasing subsequence of y, and since $y \in A$, we have $\mathrm{Card}\,(I \setminus J) \leq s$ which is the first claim of the lemma. The second claim immediately follows from the fact that the function $u \mapsto (u-s)/\sqrt{u}$ increases for $u \geq s$. Lemma 8.11 is established. □

The following is the announced concentration property.

Theorem 8.12. *Denote by m_n a median of $L_n = L_n(U_1, \ldots, U_n)$. For every $r \geq 0$,*

$$\mathbb{P}\big(\{L_n \geq m_n + r\}\big) \leq 2\,\mathrm{e}^{-r^2/4(m_n+r)}$$

and

$$\mathbb{P}\big(\{L_n \leq m_n - r\}\big) \leq 2\,\mathrm{e}^{-r^2/4m_n}.$$

In particular,

$$\mathbb{P}\big(\{|L_n - m_n| \geq r\}\big) \leq 4\,\mathrm{e}^{-r^2/8m_n}$$

for $0 \leq r \leq m_n$.

Proof. The first inequality is an immediate consequence of Theorem 4.6 combined with the preceding lemma. To establish the second inequality, we use Lemma 8.11 with $s = m_n - r$, $u = r$, to see that

$$\mathcal{D}_A^c(x) \geq \frac{r}{\sqrt{m_n}}$$

whenever $L_n(x) \geq m_n$ (recall $A = \{L_n \leq s\} = \{L_n \leq m_n - r\}$). Hence

$$\mathbb{P}\bigg(\Big\{\mathcal{D}_A^c \geq \frac{r}{\sqrt{m_n}}\Big\}\bigg) \geq \frac{1}{2}.$$

On the other hand, by Theorem 4.6 again,

$$\mathbb{P}\bigg(\Big\{\mathcal{D}_A^c \geq \frac{r}{\sqrt{m_n}}\Big\}\bigg) \leq \frac{1}{\mathbb{P}(A)}\,\mathrm{e}^{-r^2/4m_n}.$$

The required bound on $\mathbb{P}(A) = \mathbb{P}(\{L_n \leq m_n - r\})$ follows. The theorem is proved. □

We now present without proofs some further applications of the convex hull approximation referring to [Tal7] for details. We deal first with the travelling salesman problem and the minimum spanning tree.

Given a finite subset F in the unit square $[0,1]^2$ in $\mathbb{R}^2$, denote first by $L(F)$ the length of the shortest tour through F. We study the random variable $L_n = L(\{U_1, \ldots, U_n\})$ where $U_1, \ldots, U_n$ are independently uniformly distributed over $[0,1]^2$.

Theorem 8.13. *There is a numerical constant $K > 0$ such that for every $r \geq 0$,*

$$\mathbb{P}\big(\{|L_n - m_n| \geq r\}\big) \leq K\,\mathrm{e}^{-r^2/K}$$

where m_n is a median of L_n.

The proof of Theorem 8.13 is only based on the following regularity property of $L(F)$. Denote by $\mathcal{C}_k$, $k \geq 1$, the family of the 2^{2k} dyadic squares with vertices $(\ell_1 2^{-k}, \ell_2 2^{-k})$, $0 \leq \ell_1, \ell_2 \leq 2^k$. Consider $F \subset [0,1]^2$, $C \in \mathcal{C}_k$, $G \subset C$, and assume that there is a point of F within distance 2^{-k+2} of C. Then

$$L(F) \leq L(F \cup G) \leq L(F) + K\, 2^{-k} \sqrt{\mathrm{Card}(G)} \tag{8.8}$$

for some numerical constant $K > 0$.

The random miminum spanning tree satisfies a similar concentration result. A spanning tree of a finite subset $F \subset \mathbb{R}^2$ is a connected set that is a union of segments each of which joins two points of F. Its length is the sum of the lengths of these segments. We denote by $L(F)$ the length of the shortest (minimum) spanning tree of F. With respect to the travelling salesman functional, it may happen here that $L(F)$ is not monotone. We study again the random variable $L_n = L(\{U_1, \ldots, U_n\})$ where $X_1, \ldots, X_n$ are independently uniformly distributed over $[0,1]^2$.

Theorem 8.14. *There is a numerical constant $K > 0$ such that for every $r \geq 0$,*

$$\mathbb{P}\big(\{|L_n - m_n| \geq r\}\big) \leq K\, \mathrm{e}^{-r^2/K}$$

where m_n is a median of L_n.

The regularity property of L used in this theorem is now the following. Consider $F \subset [0,1]^2$ finite and $C \in \mathcal{C}_k$, $k \geq 1$. Assume that each $C' \in \mathcal{C}_{k-1}$ that is within distance 2^{-k+5} of C meets F. Then, for any $G \subset C$,

$$\big|L(F \cup G) - L(F)\big| \leq K\, 2^{-k} \sqrt{\mathrm{Card}(G)} \tag{8.9}$$

for some numerical constant $K > 0$.

Let us deal next with first passage time in percolation theory. Let $(V, \mathcal{E})$ be a graph with vertices V and edges $\mathcal{E}$. On some probability space $(\Omega, \mathcal{A}, \mathbb{P})$, let $(Y_e)_{e \in \mathcal{E}}$ be a family of non-negative independent and identically distributed random variables. Y_e represents the passage time through the edge e. Let $\mathcal{T}$ be a family of (finite) subsets of $\mathcal{E}$, and, for $T \in \mathcal{T}$, set $Y_T = \sum_{e \in T} Y_e$. If T is made of contiguous edges, Y_T represents the passage time through the path T. Set

$$Z_{\mathcal{T}} = \inf_{T \in \mathcal{T}} Y_T = \inf_{T \in \mathcal{T}} \sum_{e \in T} Y_e$$

and $D = \sup_{T \in \mathcal{T}} \mathrm{Card}(T)$, and let m be a median of $Z_{\mathcal{T}}$.

Theorem 8.15. *Assume that $0 \leq Y_e \leq 1$ almost surely. Then, for every $r \geq 0$,*

$$\mathbb{P}\big(\{|Z_{\mathcal{T}} - m| \geq r\}\big) \leq 4\, \mathrm{e}^{-r^2/4D}.$$

In particular,

$$\big|\mathbb{E}(Z_{\mathcal{T}}) - m\big| \leq 4\sqrt{\pi D} \quad \textit{and} \quad \mathrm{Var}(Z_{\mathcal{T}}) \leq 16D.$$

Proof. Changing Y_e into $-Y_e$, we may apply Corollary 4.8. Assume that V is finite. Define, for $x = (x_e)_{e \in \mathcal{E}}$,

$$F(x) = \sup_{T \in \mathcal{T}} \sum_{e \in T} x_e = \sup_{t \in \widetilde{\mathcal{T}}} \sum_{e \in V} t_e x_e$$

where $\widetilde{\mathcal{T}}$ is the collection of all $t = (t_e)_{e \in V}$ defined by

$$t_e = \mathbf{1}_{\{e \in T\}}, \quad e \in V,$$

for $T \in \mathcal{T}$. Since $\sup_{t \in \widetilde{\mathcal{T}}} \big(\sum_{e \in V} t_e^2 \big)^{1/2} \leq \sqrt{D}$, the conclusion follows from Corollary 4.8. □

When V is $\mathbb{Z}^2$ and $\mathcal{E}$ the edges connecting two adjacent points, and when $\mathcal{T} = \mathcal{T}_n$ is the set of all self-avoiding paths connecting the origin to the point $(0, n)$, H. Kesten [Ke] showed that, when $0 \leq X \leq 1$ almost surely and $\mathbb{P}(\{Y_e = 0\}) < \frac{1}{2}$ (percolation), one may reduce, in $Z_{\mathcal{T}_n}$, to paths with length less than some multiple of n. Together with this result, Theorem 8.15 indicates that for every $r \geq 0$,

$$\mathbb{P}\big(|Z_{\mathcal{T}_n} - m_n| \geq r\big) \leq K \exp\bigg(-\frac{r^2}{Kn}\bigg)$$

where m_n is a median of $\mathcal{T}_n$. This result strengthens the previous estimate by H. Kesten [Ke] which was of the order of $r/C\sqrt{n}$ in the exponent and the proof of which was based on martingale inequalities. The more delicate unbounded case is investigated in [Tal7] by means of penalty theorems.

We close this section with the assignment problem. One basic problem in this area is to study the infimum

$$\mathcal{A}_n = \inf \sum_{i=1}^{n} U_{i,\sigma(i)}$$

where the infimum is over all permutations σ of $\{1, \ldots, n\}$ and the $U_{i,j}$'s are n^2 independent random variables with the uniform distribution on $[0,1]$. It is a remarkable fact that $\mathbb{E}(\mathcal{A}_n)$ is bounded as $n \to \infty$, and actually $\mathbb{E}(\mathcal{A}_n) \leq 2$ (see [Ste] and the references therein).

M. Talagrand proved in [Tal7], again as an application of the convex hull approach of Section 4.2, that the standard deviation of $\mathcal{A}_n$ is of the order of $O(n^{-1/2})$ up to logarithmic factors.

Theorem 8.16. *Denote by m_n a median of $\mathcal{A}_n$. There is a numerical constant $K > 0$ such that for $n \geq 3$ and $r \leq \sqrt{\log n}$,*

$$\mathbb{P}\bigg(\bigg\{|\mathcal{A}_n - m_n| \geq \frac{Kr(\log n)^2}{\sqrt{n}\log\log n}\bigg\}\bigg) \leq 2\,\mathrm{e}^{-r^2}$$

while for $r \geq \sqrt{\log n}$,

$$\mathbb{P}\bigg(\bigg\{|\mathcal{A}_n - m_n| \geq \frac{Kr^3\log n}{(\log r)^2\sqrt{n}}\bigg\}\bigg) \leq 2\,\mathrm{e}^{-r^2}.$$

We refer to [Tal7] for the proofs, and to [Ste], [MD2] for further illustrations in geometric and combinatorial probabilities.

8.4 The spin glass free energy

This application concerns the limiting behavior of the spin glass free energy function at high temperature.

For $\varepsilon \in \{-1,+1\}^{\mathbb{N}}$, denote by $(\varepsilon_i)_{i\in\mathbb{N}}$ the coordinate functions that define, under the uniform product measure, independent symmetric random variables taking values ± 1. Each ε_i represents the spin of a particle i. Consider then interactions g_{ij}, $i<j$, between spins. For some parameter $\beta>0$ (that plays the role of the inverse of the temperature), the family of Gibbs measures $\widetilde{Z}_N^{-1}\mathrm{e}^{-\beta H_N}$, $N\geq 2$, on $\{-1,+1\}^N$, with Hamiltonian

$$H_N = H_N(x,\varepsilon) = -\frac{1}{\sqrt{N}}\sum_{1\leq i<j\leq N}\varepsilon_i\varepsilon_j g_{ij}$$

describe the behavior of the system with size N (at temperature $\frac{1}{\beta}$). The normalization factor $\widetilde{Z}_N$, $N\geq 2$, the so-called *partition function*, is defined by

$$\widetilde{Z}_N = \sum_{\varepsilon\in\{-1,+1\}^N}\mathrm{e}^{-\beta H_N(\varepsilon)}.$$

We will actually work with $Z_N = 2^{-N}\widetilde{Z}_N$ and use to describe Z_N the probabilistic representation

$$Z_N = Z_N^\beta = \mathbb{E}_\varepsilon\biggl(\exp\biggl(\frac{\beta}{\sqrt{N}}\sum_{1\leq i<j\leq N}\varepsilon_i\varepsilon_j g_{ij}\biggr)\biggr),$$

where $\mathbb{E}_\varepsilon$ is integration with respect to the $\varepsilon_i's$.

In the model we study, the interactions g_{ij} are random and the g_{ij}'s will be assumed independent with common standard Gaussian distribution.

Set $F_N = \log Z_N$, which defines the spin glass *free energy* (up to normalization).

Theorem 8.17. *For every* $0<\beta<1$,

$$\lim_{N\to\infty}\frac{F_N}{N} = \frac{\beta^2}{4}$$

almost surely.

The proof is based on concentration of Gaussian distribution together with the so-called second moment method, or Paley-Zygmund inequality.

Proof. We may think of F_N as function

$$F_N(x) = \log\mathbb{E}_\varepsilon\biggl(\exp\biggl(\frac{\beta}{\sqrt{N}}\sum_{1\leq i<j\leq N}\varepsilon_i\varepsilon_j x_{ij}\biggr)\biggr),\quad x=(x_{ij})_{1\leq i<j\leq N},$$

on $\mathbb{R}^n$, $n=N(N-1)/2$, equipped with the canonical Gaussian measure $\gamma=\gamma^n$. We write $\mathbb{P}$ and $\mathbb{E}$ for γ and $\int d\gamma$. It is easily seen that $\|F_N\|_{\mathrm{Lip}}\leq\beta\sqrt{(N-1)/2}$. Therefore, by (2.35) for example, for every $r\geq 0$,

$$\mathbb{P}\bigl(\bigl\{|F_N-\mathbb{E}(F_N)|\geq r\bigr\}\bigr)\leq 2\,\mathrm{e}^{-r^2/\beta^2(N-1)}. \tag{8.10}$$

Hence, by the Borel-Cantelli lemma, for any $\beta > 0$,

$$\lim_{N\to\infty} \left| \frac{F_N}{N} - \frac{\mathbb{E}(F_N)}{N} \right| = 0$$

almost surely. It therefore suffices to show that

$$\lim_{N\to\infty} \frac{\mathbb{E}(F_N)}{N} = \frac{\beta^2}{4}$$

for $0 < \beta < 1$. By Jensen's inequality, for every $N \geq 2$,

$$\mathbb{E}(F_N) = \mathbb{E}(\log Z_N) \leq \log \mathbb{E}(Z_N) = \frac{\beta^2(N-1)}{4}. \tag{8.11}$$

We have thus to prove that $\mathbb{E}(F_N)$ is actually of this order. To this task we use again the concentration result (8.10) together with the so-called second moment method.

Lemma 8.18. *For every $0 < \beta < 1$ and every N,*

$$\mathbb{E}(Z_N^2) \leq K(\beta)\, \mathrm{e}^{\beta^2(N-1)/2} = K(\beta)\, \big(\mathbb{E}(Z_N)\big)^2,$$

with

$$K(\beta) = \frac{1}{\sqrt{1-\beta^2}}.$$

Proof. By Fubini's theorem, denoting by $(\varepsilon_i')_{i\in\mathbb{N}}$ a sequence of independent symmetric Bernoulli random variables, independent of both sequences $(\varepsilon_i)_{i\in\mathbb{N}}$ and $(g_{ij})_{i,j\in\mathbb{N}}$,

$$\begin{aligned}
\mathbb{E}(Z_N^2) &= \mathbb{E}\mathbb{E}_\varepsilon\mathbb{E}_{\varepsilon'}\bigg(\exp\bigg(\frac{\beta}{\sqrt{N}} \sum_{i<j} (\varepsilon_i\varepsilon_j + \varepsilon_i'\varepsilon_j') g_{ij} \bigg)\bigg) \\
&= \mathbb{E}_\varepsilon\mathbb{E}_{\varepsilon'}\bigg(\exp\bigg(\frac{\beta^2}{2N} \sum_{i<j} (\varepsilon_i\varepsilon_j + \varepsilon_i'\varepsilon_j')^2 \bigg)\bigg).
\end{aligned}$$

Further,

$$\begin{aligned}
\sum_{i<j} (\varepsilon_i\varepsilon_j + \varepsilon_i'\varepsilon_j')^2 &= N(N-1) + 2\sum_{i<j} \varepsilon_i\varepsilon_j\varepsilon_i'\varepsilon_j' \\
&= N(N-2) + \bigg(\sum_i \varepsilon_i\varepsilon_i' \bigg)^2.
\end{aligned}$$

Now $\sum_{i=1}^N \varepsilon_i\varepsilon_i'$ has the same distribution as $\sum_{i=1}^N \varepsilon_i$. It follows therefore from the preceding that

$$\mathbb{E}(Z_N^2) \leq \mathrm{e}^{\beta^2(N-1)/2}\, \mathbb{E}_\varepsilon\bigg(\exp\bigg(\frac{\beta^2}{2N} \bigg(\sum_{i=1}^N \varepsilon_i \bigg)^2 \bigg)\bigg).$$

If g is a standard normal variable,

$$\exp\bigg(\frac{\beta^2}{2N} \bigg(\sum_{i=1}^N \varepsilon_i \bigg)^2 \bigg) = \mathbb{E}\bigg(\exp\bigg(\frac{\beta}{\sqrt{N}} \bigg(\sum_{i=1}^N \varepsilon_i \bigg) g \bigg)\bigg),$$

so that, by Fubini's theorem again,

$$\begin{aligned}\mathbb{E}_\varepsilon\bigg(\exp\Big(\frac{\beta^2}{2N}\Big(\sum_{i=1}^N\varepsilon_i\Big)^2\Big)\bigg) &= \mathbb{E}\bigg(\cosh^N\Big(\frac{\beta}{\sqrt{N}}\,g\Big)\bigg)\\ &\le \mathbb{E}\bigg(\exp\Big(\frac{\beta^2g^2}{2}\Big)\bigg)\\ &= \frac{1}{\sqrt{1-\beta^2}}\end{aligned}$$

where we used that $\cosh u \le \mathrm{e}^{-u^2/2}$. Together with (8.11), the proof of the lemma is complete. □

We turn back to the proof of the theorem. We make use of the classical Paley-Zygmund inequality

$$\mathbb{P}\big(\{Z_N \ge \tfrac{1}{2}\,\mathbb{E}(Z_N)\}\big) \ge \frac{(\mathbb{E}(Z_N))^2}{4\mathbb{E}(Z_N^2)}\,. \tag{8.12}$$

For a proof, simply note that, for every $0<\theta<1$,

$$\mathbb{E}(Z_N) \le \mathbb{E}\big(Z_N\,\mathbf{1}_{\{Z_N\ge\theta\mathbb{E}(Z_N)\}}\big) + \theta\mathbb{E}(Z_N),$$

so that, by the Cauchy-Schwarz inequality,

$$(1-\theta)\mathbb{E}(Z_N) \le \Big(\mathbb{E}(Z_N^2)\mathbb{P}\big(\{Z_N\ge\theta\mathbb{E}(Z_N)\}\big)\Big)^{1/2}.$$

Choose then $\theta = \frac{1}{2}$.

It thus follows from (8.12) and Lemma 8.18 that

$$\mathbb{P}\big(\{Z_N \ge \tfrac{1}{2}\,\mathbb{E}(Z_N)\}\big) \ge \frac{1}{4K(\beta)}\,.$$

Assume first that $r = \log(\frac{1}{2}\mathbb{E}(Z_N)) - \mathbb{E}(F_N) > 0$. Then, by (8.10) applied to this r,

$$\begin{aligned}\frac{1}{4K(\beta)} &\le \mathbb{P}\Big(\big\{\log Z_N \ge \log\big(\tfrac{1}{2}\,\mathbb{E}(Z_N)\big)\big\}\Big)\\ &\le \mathbb{P}\big(\{F_N \ge \mathbb{E}(F_N)+r\}\big)\\ &\le 2\,\mathrm{e}^{-r^2/\beta^2(N-1)}\end{aligned}$$

so that

$$r \le \beta\sqrt{N-1}\sqrt{\log 8K(\beta)}.$$

Hence, by (8.11) and the preceding, in any case,

$$\begin{aligned}\frac{\beta^2(N-1)}{4} &\ge \mathbb{E}(F_N)\\ &\ge \log\big(\tfrac{1}{2}\,\mathbb{E}(Z_N)\big) - \beta\sqrt{N-1}\sqrt{\log 8K(\beta)}\\ &\ge \frac{\beta^2(N-1)}{4} - \log 2 - \beta\sqrt{N-1}\sqrt{\log 8K(\beta)}.\end{aligned}$$

Hence $\mathbb{E}(F_N)/N \to \beta^2/4$ as $N \to \infty$ and Theorem 8.17 follows. □

It should be noted that the first part of the proof of Theorem 8.17 indicates that for any $\beta > 0$,

$$0 \le \limsup_{N\to\infty} \frac{F_N}{N} \le \frac{\beta^2}{4}$$

almost surely. We have seen in Theorem 8.17 that the upper bound $\frac{\beta^2}{4}$ is actually accurate for $0 < \beta < 1$. It may be shown that this is still the case for $\beta = 1$, but the low temperature regime $\beta \ge 1$ is of a much higher difficulty. Only few rigorous results are known, and challenging replicas conjectures of the physicists have still to be proved [Tal12], [Tal13].

Even in the high temperature regime $0 < \beta < 1$ for which Theorem 8.17 is available, a number of more precise bounds may be analyzed. For example, the proof of Theorem 8.17 displays the inequality

$$\begin{aligned}\mathbb{P}\bigg(\bigg\{\Big|F_N - \frac{\beta^2(N-1)}{4}\Big| &\ge r\beta\sqrt{N-1} \\ &+ \beta\sqrt{N-1}\sqrt{\log 8K(\beta) + \log 2}\bigg\}\bigg) \le 2\,\mathrm{e}^{-r^2}\end{aligned} \tag{8.13}$$

for $r \ge 0$. The following refines upon (8.13).

Theorem 8.19. *For each $\beta < 1$, there is a constant $K'(\beta) > 0$ such that for each N and each $r \ge 0$,*

$$\mathbb{P}\bigg(\bigg\{F_N \le \frac{\beta^2(N-1)}{4} - r\bigg\}\bigg) \le K'(\beta)\,\mathrm{e}^{-r^2/K'(\beta)}.$$

Note that by Chebyschev's inequality and Lemma 8.18, for every $s > 0$,

$$\mathbb{P}\big(\{Z_N \ge s\,\mathbb{E}(Z_N)\}\big) \le \frac{K(\beta)}{s^2}.$$

Hence, since $\log \mathbb{E}(Z_N) = \beta^2(N-1)/4$, for every $r \ge 0$,

$$\mathbb{P}\bigg(\bigg\{F_N \ge \frac{\beta^2(N-1)}{4} + r\bigg\}\bigg) \le K(\beta)\,\mathrm{e}^{-2r}. \tag{8.14}$$

Proof of Theorem 8.19. It is an important observation to note that $F_N = \log Z_N$ is convex (as a function on $\mathbb{R}^n$, $n = N(N-1)/2$). The idea is then to combine Proposition 1.6 with the following analogue of the second moment method. Indeed, we have, for every N,

$$\mathbb{E}\big(|\nabla Z_N|^2\big) \le K_1(\beta)\mathbb{E}(Z_N)^2 \tag{8.15}$$

for some constant $K_1(\beta) > 0$ only depending on β. Indeed, using the same notation as in the proof of Lemma 8.18,

$$\begin{aligned}|\nabla Z_N|^2 &= \frac{\beta^2}{N}\,\mathbb{E}_\varepsilon\mathbb{E}_{\varepsilon'}\bigg(\sum_{i<j}(\varepsilon_i\varepsilon_j\varepsilon'_i\varepsilon'_j)\exp\bigg(\frac{\beta}{\sqrt{N}}\sum_{i<j}(\varepsilon_i\varepsilon_j + \varepsilon'_i\varepsilon'_j)g_{ij}\bigg)\bigg) \\ &\le \frac{\beta^2}{2N}\,\mathbb{E}_\varepsilon\mathbb{E}_{\varepsilon'}\bigg(\Big(\sum_i \varepsilon_i\varepsilon'_i\Big)^2\exp\bigg(\frac{\beta}{\sqrt{N}}\sum_{i<j}(\varepsilon_i\varepsilon_j + \varepsilon'_i\varepsilon'_j)g_{ij}\bigg)\bigg).\end{aligned}$$

Now $u \leq \mathrm{e}^u$ for every $u \in \mathbb{R}$. Hence, arguing exactly as in the proof of Lemma 8.18,

$$\begin{aligned}\mathbb{E}\big(|\nabla Z_N|^2\big) &\leq \frac{1}{\delta}\,\mathrm{e}^{\beta^2(N-1)/2}\,\mathbb{E}_\varepsilon\bigg(\exp\bigg(\frac{\beta^2(1+\delta)}{2N}\bigg(\sum_{i=1}^N \varepsilon_i\bigg)^2\bigg)\bigg)\\ &\leq \frac{1}{\delta}\,\mathrm{e}^{\beta^2(N-1)/2}\,\frac{1}{\sqrt{1-\beta^2(1+\delta)}}\end{aligned}$$

provided that $\beta^2(1+\delta) < 1$. It then suffices to appropriately choose $\delta > 0$ depending on β so that (8.15) holds.

For $0 < \theta < 1$ and $L > 0$,

$$\begin{aligned}\mathbb{P}\big(\{Z_N \geq &\theta\,\mathbb{E}(Z_N), |\nabla Z_N| \leq L Z_N\}\big)\\ &\geq \mathbb{P}\big(\{Z_N \geq \theta\,\mathbb{E}(Z_N), |\nabla Z_N| \leq L\theta\,\mathbb{E}(Z_N)\}\big)\\ &\geq \mathbb{P}\big(\{Z_N \geq \theta\,\mathbb{E}(Z_N)\}\big) - \mathbb{P}\big(\{|\nabla Z_N| > L\theta\,\mathbb{E}(Z_N)\}\big)\\ &\geq \frac{(1-\theta)^2}{K(\beta)} - \frac{K_1(\beta)}{L^2\theta^2}\end{aligned}$$

where we have used the Paley-Zygmund inequality (8.12) and Lemma 8.18, as well as Chebyshev's inequality and (8.15) in the last step. Choose now $\theta = \frac{1}{2}$ and $L^2 = 32K(\beta)K_1(\beta)$ so that

$$\mathbb{P}\Big(\Big\{Z_N \geq \tfrac{1}{2}\,\mathbb{E}(Z_N), |\nabla Z_N| \leq \sqrt{32K(\beta)K_1(\beta)}\,Z_N\Big\}\Big) \geq \frac{1}{8K(\beta)}\,.$$

In terms of $F_N = \log Z_N$,

$$\mathbb{P}\Big(\Big\{F_N \geq \log\big(\tfrac{1}{2}\,\mathbb{E}(Z_N)\big), |\nabla F_N| \leq \sqrt{32K(\beta)K_1(\beta)}\Big\}\Big) \geq \frac{1}{8K(\beta)}\,.$$

We then make use of the deviation inequality of Proposition 1.6 for the convex function F_N with respect to the Gaussian measure γ on $\mathbb{R}^n$ to get that for every $r \geq 0$,

$$\mathbb{P}\bigg(\bigg\{F_N \leq \frac{\beta^2(N-1)}{4} - \log 2 - \sqrt{32K(\beta)K_1(\beta)}\bigg(r + \sqrt{2\log\frac{1}{8K(\beta)}}\bigg)\bigg\}\bigg) \leq \mathrm{e}^{-r^2/2}.$$

The conclusion of Theorem 8.19 immediately follows. □

8.5 Concentration of random matrices

Let $M^n = (M_{ij}^n)_{1\leq i,j\leq n}$ be an $n \times n$ real symmetric matrix. We are interested here in the case the entries M_{ij}^n of M^n are real-valued random variables on some probability space $(\Omega, \mathcal{A}, \mathbb{P})$, and we study the concentration properties of functionals of the eigenvalues $\lambda_1, \ldots, \lambda_n$ of M^n. The motivation for such investigation comes from Wigner's theorem (cf. e.g. [Me], [Hi-P1]) asserting that whenever the random variables M_{ij}^n, $1 \leq i \leq j \leq n$, are independent with finite moments and $\mathbb{E}(M_{ij}^n) = 0$, $\mathbb{E}((M_{ij}^n)^2) = \frac{1}{n}$, $1 \leq i < j \leq n$, then the empirical distribution $\frac{1}{n}\sum_{i=1}^n \delta_{\lambda_i}$ converges almost surely towards the semicircular law with density $(2\pi)^{-1}\sqrt{4-x^2}$ on $[-2,+2]$. The concentration properties presented here are milder results, which, however, hold for any fixed n. Moreover, with respect to the exponential large deviation bounds (with rate n^2) on Wigner's limit (cf. [BA-G], [Hi-P2]), the results here concern rather general families of underlying distributions.

The first observation allows simple use of the contraction property. Denote by $\mathcal{M}^n$ the real $n \times n$ matrices, and by $\mathcal{M}_s^n$ its symmetric part. Each element M of $\mathcal{M}_s^n$ has a unique list of eigenvalues $\lambda = \lambda(M) = (\lambda_1, \dots, \lambda_n)$ listed in increasing order according to multiplicity in the simplex

$$\mathcal{S}^n = \{\lambda_1 \le \cdots \le \lambda_n; \lambda_i \in \mathbb{R}, i = 1, \dots, n\}.$$

Equip $\mathcal{M}_s^n$ with the Hilbert-Schmidt norm

$$\|M\|_2^2 = \operatorname{Tr}(M\,{}^tM) = \sum_{i,j=1}^n M_{ij}^2$$

and $\mathcal{S}^n$ with the Euclidean norm $|\lambda| = \left(\sum_{i=1}^n \lambda_i^2\right)^{1/2}$. The following statement is a classical result (cf. e.g. [H-J], [Si]).

Proposition 8.20. *The map $\varphi : \mathcal{M}_s^n \to \mathcal{S}^n$ which associates to each real symmetric matrix its ordered list of eigenvalues is 1-Lipschitz.*

The following is then a direct consequence of the contraction principle (Proposition 1.2) and the concentration inequalities for Lipschitz functions. We use a probabilistic formulation. If F is a function on $\mathcal{S}^n$, set $\widetilde{F} = F \circ \varphi$ on $\mathcal{M}_s^n$. A typical example is given by

$$\widetilde{F}(M) = \operatorname{Tr} f(M) = \sum_{i=1}^n f(\lambda_i) \tag{8.16}$$

for some function $f : \mathbb{R} \to \mathbb{R}$.

Corollary 8.21. *Let M be a symmetric random matrix, and denote by P the distribution of its entries M_{ij}, $1 \le i \le j \le n$, on $\mathcal{M}_s^n$. Denote as usual by α_P the concentration function of P. For any 1-Lipschitz function $F : \mathcal{S}^n \to \mathbb{R}$ and any $r > 0$,*

$$\mathbb{P}\Big(\big\{|\widetilde{F}(M) - m| \ge r\big\}\Big) \le 2\alpha_P(r)$$

where m is a median of $\widetilde{F}(M)$.

Under appropriate integrability properties, the same inequality holds with the mean replaced by the median by the results of Section 1.3.

Assume for example, as in Wigner's theorem, that the M_{ij}'s are independent centered real Gaussian variables with variance $\frac{1}{n}$. We then get from Corollary 8.21 that if F is 1-Lipschitz, for any $r \ge 0$,

$$\mathbb{P}\Big(\big\{|\widetilde{F}(M^n) - m| \ge r\big\}\Big) \le 2\,\mathrm{e}^{-nr^2/2}$$

where m is either the mean or a median of $\widetilde{F}(M)$. In particular, if $f : \mathbb{R} \to \mathbb{R}$ is 1-Lipschitz,

$$\mathbb{P}\Big(\big\{|\tfrac{1}{n}\operatorname{Tr} f(M) - m| \ge r\big\}\Big) \le 2\,\mathrm{e}^{-n^2r^2/2}, \quad r \ge 0.$$

Another example of a Lipschitz function is given by

$$\widetilde{F}(M) = \max_{1\leq i\leq n} \lambda_i = \lambda_n \tag{8.17}$$

for which we get that

$$\mathbb{P}\big(\{|\max_{1\leq i\leq n} \lambda_i - m| \geq r\}\big) \leq 2\,\mathrm{e}^{-nr^2/2}, \quad r \geq 0,$$

where m is either the mean or the median of $\max_{1\leq i\leq n}\lambda_i$. As for the longest increasing subsequence however, much more precise fluctuation results are available in this case [So].

Similarly, we may consider the case where the entries M_{ij}'s of $M \in \mathcal{M}_s^n$ are independent bounded random variables for which the results of Chapter 4 provide a number of concentration results for convex functionals. To this task, the following classical result (cf. [Davi], [H-J) is of use. A function $F : \mathbb{R}^n \to \mathbb{R}$ is said to be symmetric if it is invariant under permutations.

Proposition 8.22. *Let $F : \mathbb{R}^n \to \mathbb{R}$ be convex and symmetric. Then $\widetilde{F} = F \circ \varphi$ on $\mathcal{M}_s^n$ is a (real orthogonally invariant) convex matrix function.*

The next corollary, consequence of Corollary 4.10, is then the analogue of Corollary 8.21. Note that if $f : \mathbb{R} \to \mathbb{R}$ is convex, then Proposition 8.22 applies to $\mathrm{Tr}\, f(M)$ as defined in (8.16) as well as to $\max_{1\leq i\leq n}\lambda_i$ of (8.17).

Corollary 8.23. *Let M be a symmetric random matrix, and assume the entries M_{ij}, $1 \leq i \leq j \leq n$, are independent random variables bounded (by 1). Then, for any convex symmetric 1-Lipschitz function $F : \mathcal{S}^n \to \mathbb{R}$, and any $r > 0$,*

$$\mathbb{P}\Big(\big\{|\widetilde{F}(M) - m| \geq r\big\}\Big) \leq 4\,\mathrm{e}^{-r^2/4}$$

where m is a median of $\widetilde{F}(M)$.

Complex entries are treated similarly. The structure of the arguments is sufficiently general to cover in the same way symmetric inhomogeneous matrices with entries $a_{ij}M_{ij}^n$ where the a_{ij} are fixed non-random coefficients. This provides concentration for a number of families of random matrices including diluted, band or Wishart matrices. We refer to [G-Zei] for details.

An interesting by-product of Proposition 8.22 is that whenever $U : \mathbb{R}^n \to \mathbb{R}$ is symmetric and strictly convex in the sense that for some $c > 0$, $\mathrm{Hess}\, U(x) \geq c\,\mathrm{Id}$ uniformly in $x \in \mathbb{R}^n$, then the probability measure μ on $\mathcal{M}_s^n$ defined by

$$d\mu(M) = \frac{1}{Z}\,\mathrm{e}^{-\widetilde{U}(M)}dM \tag{8.18}$$

(where dM is Lebesgue measure on $\mathcal{M}_s^n$) is also strictly convex in the same sense. In particular, it satisfies the Gaussian like isoperimetric inequality of Theorem 2.7 and thus has normal concentration independently of the dimension n. By Theorem 5.2, the measure μ also satisfies a logarithmic Sobolev inequality. While in the Gaussian case $U(x) = |x|^2$, the preceding distribution μ on $\mathcal{M}_s^n$ may be represented by a Wigner matrix with independent Gaussian entries, a similar simple description of the ensemble in terms of the matrix entries should not be expected in the general case, the entries being possibly dependent as random variables. For p-convex potentials, $p \geq 2$, see [Bl2].

Notes and Remarks

The topics selected in this chapter only reflect a few of the applications of the concentration of measure phenomenon. Geometric and topological applications were already described in Chapter 3, together with the historical application to Euclidean sections of convex bodies. Applications to sums of independent random vectors and supremum of empirical processes are the subject of Chapter 7.

Concentration for harmonic measures as described in the first section is taken from the work [S-S2] by G. Schechtman and M. Schmuckenschläger, to which the reader is referred for further details and alternative, more analytic, proofs, in particular of Theorem 8.5. Recently, F. Barthe [Bar2] was able to show that the harmonic measures σ_x^n satisfy a Poincaré inequality with a uniform constant of the order of $\frac{1}{n}$ thus recovering exponential concentration from Section 3.1. The corresponding conclusion for the logarithmic Sobolev inequality is so far open.

Concentration for independent permutations started with the note [Mau1] by B. Maurey based on the martingale method. The question of the convex envelope is addressed in [Tal7]. The extension and result presented in Section 8.2 are taken from the recent contribution [MD3] by C. McDiarmid motivated by randomized methods for graph coloring (cf. [M-R]).

The memoir [Tal7] by M. Talagrand contains a number of applications to discrete and geometric probabilities from which the few topics presented in Section 8.3 are taken. Complete expositions in the framework of probability theory for algorithmic discrete mathematics and combinatorial optimization are the notes [MD2] by C. McDiarmid and [Ste] by M. Steele. For sharp constants in Theorem 8.12 via the entropic method, see [Bo-L-M]. That concentration methods do not provide tight results for the longest increasing subsequence problem is observed in [Deu-Z]. Far reaching fluctuation results on the distribution of the longest increasing subsequence have been obtained by combinatorial and analytic methods in [B-D-J].

The application of concentration for Gaussian measures to the spin glass free energy is due to M. Talagrand [Tal7], [Tal12], [Tal13], from which the results of Section 8.4 are taken. Theorem 8.17 goes back to [A-L-R] with a proof based on moment expansions (see also [C-N] for a stochastic calculus proof).

A general introduction to random matrices is the reference [Me]. Concentration of the spectral measure of random matrices was investigated recently by A. Guionnet and O. Zeitouni in [G-Zei] for the trace functional, and by G. Blower [Bl2] for measures with p-convex potentials. Concentration in free probability is alluded to in [V-D-N].

REFERENCES

[A-M-S] S. AIDA, T. MASUDA, I. SHIGEKAWA. Logarithmic Sobolev inequalities and exponential integrability. J. Funct. Anal. 126, 83–101 (1994).

[A-S] S. AIDA, D. STROOCK. Moment estimates derived from Poincaré and logarithmic Sobolev inequalities. Math. Res. Lett. 1, 75–86 (1994).

[A-L-R] M. AIZENMAN, J. L. LEBOWITZ, D. RUELLE. Some rigorous results on the Sherrington-Kirkpatrick spin glass model. Comm. Math. Phys. 112, 3–20 (1987).

[Ale] S. ALESKER. Localization technique on the sphere and the Gromov-Milman theorem on the concentration phenomenon on uniformly convex sphere. Convex Geometric Analysis (Berkeley 1996). Math. Sci. Res. Inst. Publ. 34. Cambridge Univ. Press (1999).

[Alo] N. ALON. Eigenvalues and expanders. Combinatorica 6, 83–96 (1986).

[Al-M] N. ALON, V. MILMAN. λ_1, isoperimetric inequalities for graphs ans superconcentrators. J. Combin. Theory, Ser. B, 38, 78–88 (1985).

[Am-M] D. AMIR, V. MILMAN. Unconditional and symmetric sets in n-dimensional normed spaces. Israel J. Math. 37, 3–20 (1980).

[An] C. ANÉ ET AL. Sur les inégalités de Sobolev logarithmiques. Panoramas et Synthèses, vol. 10. Soc. Math. de France (2000).

[A-L] C. ANÉ, M. LEDOUX. On logarithmic Sobolev inequalities for continuous time random walks on graphs. Theory Relat. Fields 116, 573–602 (2000).

[A-B-P] M. ANTILLA, K. BALL, I. PERISSINAKI. The central limit theorem for convex bodies (2000).

[AR-V] J. ARIAS-DE-REYNA, R. VILLA. The uniform concentration of measure phenomenon in ℓ_p^n, $(1 \le p \le 2)$. Geometric Aspects of Functional Analysis, Israel Seminar 1996-2000. Lecture Notes in Math. 1745, 13–18 (2000). Springer.

[AR-B-V] J. ARIAS-DE-REYNA, K. BALL, R. VILLA. Concentration of the distance in finite dimensional normed spaces. Mathematica 45, 245–252 (1998).

[A-V] R. AZENCOTT, N. VAYATIS. Refined exponential rates in Vapnik-Chervonenkis inequalities. C. R. Acad. Sci. Paris 332, 563–568 (2001).

[Az] K. AZUMA. Weighted sums of certain dependent random variables. Tohoku Math. J. 19, 357–367 (1967).

[B-T] A. BAERENSTEIN II, B. A. TAYLOR. Spherical rearrangements, subharmonic functions and $*$-functions in n-space. Duke Math. J. 43, 245–268 (1976).

[B-D-J] J. BAIK, P. A. DEIFT, K. JOHANSSON. On the distribution of the longest increasing subsequence in a random permutation. J. Amer. Math. Soc. 12, 1119–1178 (1999).

[Bak1] D. BAKRY. L'hypercontractivité et son utilisation en théorie des semigroupes. Ecole d'Eté de Probabilités de St-Flour. Lecture Notes in Math. 1581, 1–114 (1994). Springer.

[Bak2] D. BAKRY. On Sobolev and logarithmic Sobolev inequalities for Markov semigroups. New trends in Stochastic Analysis. 43–75 (1997). World Scientific.

[B-E] D. BAKRY, M. EMERY. Diffusions hypercontractives. Séminaire de Probabilités XIX. Lecture Notes in Math. 1123, 177–206 (1985). Springer.

[Ba-L] D. BAKRY, M. LEDOUX. Lévy-Gromov's isoperimetric inequality for an infinite dimensional diffusion generator. Invent. math. 123, 259–281 (1996).

[Bal] K. BALL. An elementary introduction to modern convex geometry. Flavors of Geometry. Math. Sci. Research Inst. Publ. 31. Cambridge Univ. Press (1997).

[B-B-M] A. BARRON, L. BIRGÉ, P. MASSART. Risk bounds for model selection via penalization. Probab. Theory Relat. Fields 113, 301–413 (1999).

[Bar1] F. BARTHE. Levels of concentration between exponential and Gaussian (2000).

[Bar2] F. BARTHE. Poincaré inequalities for harmonic measures on spheres (2000).

[BA-G] G. BEN AROUS, A. GUIONNET. Large deviations for Wigner's law and Voiculescu's non-commutative entropy. Probab. Theory Relat. Fields 108, 517–542 (1997).

[Benn] G. BENNETT. Probability inequalities for sums of independent random variables. J. Amer. Statist. Assoc. 57, 33–45 (1962).

[Beny] Y. BENYAMINI. Two point symmetrization, the isoperimetric inequality on the sphere and some applications. Longhorn Notes, Functional Analysis Seminar 1983-84, 53–76, University of Texas.

[B-M1] L. BIRGÉ, P. MASSART. From model selection to adaptive estimation. Festschrift for Lucien LeCam: Research papers in Probability and Statistics (D. Pollard, E. Torgersen and G. Yang, eds.) 55–87 (1997). Springer.

[B-M2] L. BIRGÉ, P. MASSART. Minimum contrast estimators on sieves: exponential bounds and rates of convergence. Bernoulli 4, 329–375 (1998).

[Bl1] G. BLOWER. The Gaussian isoperimetric inequality and transportation (1999). Positivity, to appear.

[Bl2] G. BLOWER. Almost sure weak convergence for the generalized orthogonal ensemble (2000).

[Bob1] S. BOBKOV. On Gross' and Talagrand's inequalities on the discrete cube. Vestnik of Syktyvkar University, 1, 12–19 (1995) (in Russian).

[Bob2] S. BOBKOV. Isoperimetric inequalities for distributions of exponential type. Ann. Probab. 22, 978–994 (1994).

[Bob3] S. BOBKOV. A functional form of the isoperimetric inequality for the Gaussian measure. J. Funct. Anal. 135, 39–49 (1996).

[Bob4] S. BOBKOV. An isoperimetric inequality on the discrete cube and an elementary proof of the isoperimetric inequality in Gauss space. Ann. Probab. 25, 206–214 (1997).

[Bob5] S. BOBKOV. Isoperimetric and analytic inequalities for log-concave probability measures. Ann. Probab., 27, 1903–1921 (1999).

[Bob6] S. BOBKOV. On deviations from medians (1999).

[Bob7] S. BOBKOV. Remarks on the growth of L^p-norms on polynomials. Geometric Aspects of Functional Analysis, Israel Seminar 1996–2000. Lecture Notes in Math. 1745, 27–35 (2000). Springer.

[B-G-L] S. BOBKOV, I. GENTIL, M. LEDOUX. Hypercontractivity of Hamilton-Jacobi equations (2000). J. Math. Pures Appl., to appear.

[B-G] S. BOBKOV, F. GÖTZE. Exponential integrability and transportation cost related to logarithmic Sobolev inequalities. J. Funct. Anal. 163, 1–28 (1999).

[B-G-H] S. BOBKOV, F. GÖTZE, C. HOUDRÉ. On Gaussian and Bernoulli covariance representations. Bernoulli 7, 439–452 (2001).

[B-H-T] S. BOBKOV, C. HOUDRÉ, P. TETALI. λ_∞, vertex isoperimetry and concentration. Combinatorica 20, 153–172 (2000).

[Bo-L1] S. BOBKOV, M. LEDOUX. Poincaré's inequalities and Talagrand's concentration phenomenon for the exponential measure. Probab. Theory Relat. Fields 107, 383–400 (1997).

[Bo-L2] S. BOBKOV, M. LEDOUX. On modified logarithmic Sobolev inequalities for Bernoulli and Poisson measures. J. Funct. Anal. 156, 347–365 (1998).

[Bo-L3] S. BOBKOV, M. LEDOUX. From Brunn-Minkowki to Brascamp-Lieb and to logarithmic Sobolev inequalities. Geom. funct. anal. 10, 1028–1052 (2000).

[B-H] TH. BODINEAU, B. HELFFER. On Log-Sobolev inequalities for unbounded spin systems. J. Funct. Anal. 166, 168–178 (1999).

[Bog] V. BOGACHEV. Gaussian measures. Amer. Math. Soc. (1998).

[Bor1] C. BORELL. Convex measures on locally convex spaces. Ark. Math. 12, 239–252 (1974).

[Bor2] C. BORELL. The Brunn-Minkowski inequality in Gauss space. Invent. math. 30, 207–216 (1975).

[Bo-L-M] S. BOUCHERON, G. LUGOSI, P. MASSART. A sharp concentration inequality with applications. Random Structures and Algorithms 16, 277–292 (2000).

[Bou] J. BOURGAIN. On the distribution of polynomials on high dimensional convex sets. Geometric Aspects of Functional Analysis (Israel Seminar, 1989–90). Lecture Notes in Math. 1469, 127–137 (1991). Springer.

[B-L-M] J. BOURGAIN, J. LINDENSTRAUSS, V. D. MILMAN. Approximation of zonoids by zonotopes. Acta Math. 162, 73–141 (1989).

[Bre] Y. BRENIER. Polar factorization and monotone rearrangement of vector-valued functions. Comm. Pure Appl. Math. 44, 375–417 (1991).

[Bro] R. BROOKS. On the spectrum of non-compact manifolds with finite volume. Math. Z. 187, 425–437 (1984).

[B-Z] Y. D. BURAGO, V. A. ZALGALLER. Geometric inequalities. First Edition (in Russian): Nauka (1980). Second Edition (English transl.): Springer (1988).

[Ca] O. CATONI. Free energy estimates and deviation inequalities (2000). Ann. Inst. H. Poincaré, to appear.

[Cha1] I. CHAVEL. Eigenvalues in Riemannian geometry. Academic Press (1984).

[Cha2] I. CHAVEL. Riemannian geometry—A modern introduction. Cambridge Univ. Press (1993).

[Chee] J. CHEEGER. A lower bound for the smallest eigenvalue of the Laplacian. Problems in Analysis. Princeton, 195–199 (1970).

[C-E] J. CHEEGER, D. EBIN. Comparison theorems in Riemannian geometry. North-Holland (1975).

[Chen] S.-Y. CHENG. Eigenvalue comparison theorems and its geometric applications. Math. Z. 143, 289–297 (1975).

[C-N] F. COMETS, J. NEVEU. The Sherrington-Kirkpatrick model of spin glasses and stochastic calculus: the high temperature case. Comm. Math. Phys. 166, 549–564 (1995).

[CE] D. CORDERO-ERAUSQUIN. Some applications of mass transport to Gaussian type inequalities (2000). Archiv der Math., to appear.

[CE-MC-S] D. CORDERO-ERAUSQUIN, R. MCCANN, M. SCHMUCKENSCHLÄGER. A Riemannian interpolation inequality à la Borell, Brascamp and Lieb (2000). Invent. Math., to appear.

[DG] S. DAS GUPTA. Brunn-Minkowski inequality and its aftermath. J. Multivariate Anal. 10, 296–318 (1980).

[Davi] C. DAVIS. All convex invariant functions of hermitian matrices. Archiv der Math. 8, 276–278 (1957).

[Davie1] E. B. DAVIES. One-parameter semigroups. Academic Press (1980).

[Davie2] E. B. DAVIES. Heat kernel and spectral theory. Cambridge Univ. Press (1989).

[Da-S] E. B. DAVIES, B. SIMON. Ultracontractivity and the heat kernel for Schrödinger operators and Dirichlet Laplacians. J. Funct. Anal. 59, 335–395 (1984).

[De] A. DEMBO. Information inequalities and concentration of measure. Ann. Probability 25, 927–939 (1997).

[Dem-Z1] A. DEMBO, O. ZEITOUNI. Large deviations techniques and applications. First Edition: Jones & Bartlett Publishers (1993). Second Edition: Springer (1998).

[Dem-Z2] A. DEMBO, O. ZEITOUNI. Transportation approach to some concentration inequalities in product spaces. Elect. Comm. in Probab. 1, 83–90 (1996).

[De-S] J.-D. DEUSCHEL, D. STROOCK. Large deviations. Academic Press (1989).

[Deu-Z] J.-D. DEUSCHEL, O. ZEITOUNI. On increasing subsequences of i.i.d. samples. Combinatorics, Probability and Computing 8, 247–263 (1999).

[D-SC] P. DIACONIS, L. SALOFF-COSTE. Logarithmic Sobolev inequalities for finite Markov chains. Ann. Appl. Prob. 6, 695–750 (1996).

[Dud] R. M. DUDLEY. Real analysis and probability. Chapman & Hall (1989).

[Du-S] N. DUNFORD, J. SCHWARTZ. Linear Operators. Part II: Spectral Theory. Interscience (1963).

[Dur] R. DURRETT. Brownian motion and martingales in analysis. Wordsworth (1984).

[Dv] A. DVORETZKY. Some results on convex bodies and Banach spaces. Proc. Symp. on Linear Spaces, Jerusalem, 123–160 (1961).

[D-R] A. DVORETZKY, C. A. ROGERS. Absolute and unconditional convergence in normed linear spaces. Proc. Nat. Acad. Sci. U.S.A. 36, 192–197 (1950).

[Eh] A. EHRHARD. Symétrisation dans l'espace de Gauss. Math. Scand. 53, 281–301 (1983).

[Ev] L. C. EVANS. Partial differential equations. Amer. Math. Soc. (1997).

[Fe] X. FERNIQUE. Fonctions aléatoires gaussiennes, vecteurs aléatoires gaussiens. Les publications CRM, Montréal (1997).

[F-L-M] T. FIGIEL, J. LINDENSTRAUSS, V. D. MILMAN. The dimensions of almost spherical sections of convex bodies. Acta Math. 139, 52–94 (1977).

[F-F] P. Frankl, Z. Füredi. A short proof for a theorem of Harper about Hamming spheres. Discrete Math. 34, 311-313 (1981).

[Fr] A. Friedman. Partial differential equations. Holt, Rinehart, Winston (1969).

[F-O-T] M. Fukushima, Y. Oshima, M. Takeda. Dirichlet forms and symmetric Markov processes. De Gruyter (1994).

[G-H-L] S. Gallot, D. Hulin, J. Lafontaine. Riemannian Geometry. Second Edition. Springer (1990).

[Gi-M1] A. Giannopoulos, V. Milman. Euclidean structure in finite dimensional normed spaces. The Handbook in the Geometry of Banach Spaces. Elsevier (2001).

[Gi-M2] A. Giannopoulos, V. Milman. Concentration property on probability spaces. Advances in Math. 156 (2000).

[Grom1] M. Gromov. Paul Lévy's isoperimetric inequality. Preprint I.H.E.S. (1980).

[Grom2] M. Gromov. Metric structures for Riemannian and non-Riemannian spaces. Birkhäuser (1998).

[Grom3] M. Gromov. Spaces and question. Proc. of "Visions in Mathematics—Towards 2000". GAFA 2000, Special volume, 118–161 (2000).

[Gr-M1] M. Gromov, V. D. Milman. A topological application of the isoperimetric inequality. Amer. J. Math. 105, 843–854 (1983).

[Gr-M2] M. Gromov, V. D. Milman. Generalization of the spherical isoperimetric inequality to uniformly convex Banach spaces. Compositio Math. 62, 263–282 (1987).

[Gros1] L. Gross. Logarithmic Sobolev inequalities. Amer. J. Math. 97, 1061–1083 (1975).

[Gros2] L. Gross. Logarithmic Sobolev inequalities and contractive properties of semigroups. Dirichlet Forms, Varenna 1992. Lect. Notes in Math. 1563, 54–88 (1993). Springer.

[G-R] L. Gross, O. Rothaus. Herbst inequalities for supercontractive semigroups. J. Math. Kyoto Univ. 38, 295–318 (1998).

[Gu] O. Guédon. Kahane-Khinchine type inequalities for negative exponent. Mathematica 46, 165–173 (1999).

[G-Zeg] A. Guionnet, B. Zegarlinski. Lectures on logarithmic Sobolev inequalities (2000). Séminaire de Probabilités. Lecture Notes in Math., to appear. Springer.

[G-Zei] A. Guionnet, O. Zeitouni. Concentration of the spectral measure for large matrices. Elect. Comm. Probab. 5, 119-136 (2000).

[Hi-P1] F. Hiai, D. Petz. The semicircle law, free random variables and entropy. Amer. Math. Soc. (2000).

[Hi-P2] F. Hiai, D. Petz. A large deviation theorem for the empirical distribution of random unitary matrices. Ann. Inst. H. Poincaré 36, 71–85 (2000).

[Ha] L. H. Harper. Optimal numbering and isoperimetric problems on graphs. J. Comb. Th. 1, 385–393 (1966).

[Hoe] W. J. Hoeffding. Probability inequalities for sums of bounded random variables. J. Amer. Statist. Assoc. 58, 713–721 (1963).

[H-J] R. A. Horn, C. Johnson. Matrix analysis. Cambridge Univ. Press (1985).

[Hou] C. Houdré. Remarks on deviation inequalities for functions of infinitely divisible random vectors (2000).

[Ho-P] C. Houdré, N. Privault. Deviation inequalities in infinite dimensions: An approach via covariance representations (2001).

[H-T] C. Houdré, P. Tetali. Concentration of measure for products of Markov kernels via functional inequalities. Combin, Probab. Comput. 10, 1–28 (2001).

[H-PA-S] C. Houdré, V. Pérez-Abreu, D. Surgailis. Interpolation, correlation identities and inequalities for infinitely divisible variables. J. Fourier Anal. Appl. 4, 651–668 (1998).

[Hs] E. P. Hsu. Analysis on path and loop spaces. Probability theory and applications (Princeton, NJ, 1996), 277–347, IAS/Park City Math. Ser. 6. Amer. Math. Soc. (1999).

[I-S-T] I. A. Ibragimov, V. N. Sudakov, B. S. Tsirel'son. Norms of Gaussian sample functions. Proceedings of the third Japan-USSR Symposium on Probability Theory. Lecture Notes in Math. 550, 20–41 (1976). Springer.

[I-W] N. Ikeda, S. Watanabe. Stochastic differential equations and diffusion processes. North-Holland (1989).

[J-L] W. B. Johnson, J. Lindenstrauss. The Handbook in the Geometry of Banach Spaces (Editors). Elsevier (2001).

[J-S1] W. B. JOHNSON, G. SCHECHTMAN. Remarks on Talagrand's deviation inequality for Rademacher functions. Functional Analysis Seminar 1987–89, University of Texas. Lecture Notes in Math. 1470, 72–77 (1991). Springer.

[J-S2] W. B. JOHNSON, G. SCHECHTMAN. Embedding ℓ_p^m into ℓ_1^n. Acta Math. 149, 71–85 (1982).

[J-S3] W. B. JOHNSON, G. SCHECHTMAN. Finite dimensional subspaces of L_p. The Handbook in the Geometry of Banach Spaces. Elsevier (2001).

[Ka] J.-P. KAHANE. Some random series of functions. First Edition: Heath Math. Monographs (1968). Second Edition: Cambridge Univ. Press (1985).

[K-L-S] R. KANNAN, L. LOVÁSZ, M. SIMONOVITS. Isoperimetric problems for convex bodies and a localization lemma. Discrete Comput. Geom. 13, 541–559 (1995).

[Ke] H. KESTEN. On the speed of convergence in first-passage percolation. Ann. Appl. Probab. 3, 2 96–338 (1993).

[Kl] C. KLAASSEN. On an inequality of Chernoff. Ann. Probab. 13, 966–974 (1985).

[Ko-S] A. KORZENIOWSKI, D. STROOCK. An example in the theory of hypercontractive semigroups. Proc. Amer. Math. Soc. 94, 87–90 (1985).

[Kw] S. KWAPIEŃ. A theorem on the Rademacher series with vector valued coefficients. Probability in Banach Spaces, Oberwolfach 1975. Lecture Notes in Math. 526, 157–158 (1976). Springer.

[Kw-S] S. KWAPIEŃ, J. SZULGA. Hypercontraction methods in moment inequalities for series of independent random variables in normed spaces. Ann. Probab. 19, 369–379 (1991).

[K-W] S. KWAPIEŃ, W. A. WOYCZYŃSKI. Random series and stochastic integrals: Single and multiple. Birkhäuser (1992).

[K-L-O] S. KWAPIEŃ, R. LATAŁA, K. OLESZKIEWICZ. Comparison of moments of sums of independent random variables and differential inequalities. J. Funct. Anal. 136, 258–268 (1996).

[L-O] R. LATAŁA, K. OLESKIEWICZ. Between Sobolev and Poincaré. Geometric Aspects of Functional Analysis, Israel Seminar 1996–2000. Lecture Notes in Math. 1745, 147–168 (2000). Springer.

[Le1] M. LEDOUX. A heat semigroup approach to concentration on the sphere and on a compact Riemannian manifold. Geom. funct. anal. 2, 221–224 (1992).

[Le2] M. LEDOUX. Remarks on logarithmic Sobolev constants, exponential integrability and bounds on the diameter. J. Math. Kyoto Univ. 35, 211–220 (1995).

[Le3] M. LEDOUX. Isoperimetry and Gaussian Analysis. Ecole d'Eté de Probabilités de St.-Flour 1994. Lecture Notes in Math. 1648, 165–294 (1996). Springer.

[Le4] M. LEDOUX. On Talagrand's deviation inequalities for product measures. ESAIM Prob. & Stat. 1, 63–87 (1996).

[Le5] M. LEDOUX. Concentration of measure and logarithmic Sobolev inequalities. Séminaire de Probabilités XXXIII. Lecture Notes in Math. 1709, 120–216 (1999). Springer.

[Le6] M. LEDOUX. The geometry of Markov diffusion generators. Ann. Fac. Sci. Toulouse IX, 305–366 (2000).

[Le7] M. LEDOUX. Logarithmic Sobolev inequalities for unbounded spin systems revisited. Séminaire de Probabilités XXXV. Lecture Notes in Math. 1755, 167–194 (2000). Springer.

[Le-T] M. LEDOUX, M. TALAGRAND. Probability in Banach spaces (Isoperimetry and processes). Springer (1991).

[Le-Z] M. LEDOUX, J. ZINN. Probabilistic limit theorem in the setting of Banach spaces. The Handbook in the Geometry of Banach Spaces. Elsevier (2001).

[Lé] P. LÉVY. Problèmes concrets d'analyse fonctionnelle. Gauthier-Villars (1951).

[Li] M. A. LIFSHITS. Gaussian random functions. Kluwer (1995).

[Li-T] J. LINDENSTRAUSS, L. TZAFRIRI. Classical Banach spaces II. Springer (1979).

[L-S] L. LOVÁSZ, M. SIMONOVITS. Random walks in a convex body and an improved volume algorithm. Random Structures and Algorithms 4, 369–412 (1993).

[L-Z] T. LYONS, W. ZHENG. A crossing estimate for the canonical process on a Dirichlet space and tightness result. Colloque Paul Lévy, Astérisque 157-158, 249–272 (1988).

[Ly] T. LYONS. Random thoughts on reversible potential theory. Summer School in Potentiel Theory, Joensuu 1990. Publications in Sciences 26, 71–114. University of Joensuu.

[MC] R. J. MCCANN. Existence and uniqueness of monotone measure-preserving maps. Duke Math. J. 80, 309–323 (1995).

[MD1] C. McDiarmid. On the method of bounded differences. Surveys in Combinatorics. London Math. Soc. Lecture Notes 141, 148–188 (1989). Cambridge Univ. Press.

[MD2] C. McDiarmid. Concentration. Probabilistic methods for algorithmic discrete mathematics, 195–248. Springer (1998).

[MD3] C. McDiarmid. Concentration for independent permutations (2000).

[MK] H. P. McKean. Geometry of differential space. Ann. Probability 1, 197–206 (1973).

[Mal] P. Malliavin. Stochastic analysis. Springer (1997).

[Mar1] K. Marton. Bounding $\bar{d}$-distance by information divergence: a method to prove measure concentration. Ann. Probab. 24, 857–866 (1996).

[Mar2] K. Marton. A measure concentration inequality for contracting Markov chains. Geom. funct. anal. 6, 556–571 (1997).

[Mar3] K. Marton. Measure concentration for a class of random processes. Probab. Theory Relat. Fields 110, 427–439 (1998).

[Mar4] K. Marton. On a measure concentration of Talagrand for dependent random variables (1998).

[Mas1] P. Massart. About the constants in Talagrand's deviation inequalities for empirical processes (1998). Ann. Probab. 28, 863–884 (2000).

[Mas2] P. Massart. Some applications of concentration inequalities to statistics. Ann. Fac. Sci. Toulouse IX, 245–303 (2000).

[Mau1] B. Maurey. Constructions de suites symétriques. C. R. Acad. Sci. Paris 288, 679–681 (1979).

[Mau2] B. Maurey. Some deviations inequalities. Geom. funct. anal. 1, 188–197 (1991).

[Me] M. L. Mehta. Random matrices. Academic Press (1991).

[M-P] M. Meyer, A. Pajor. Sections of the unit ball of L_p^n. J. Funct. Anal. 80, 109–123 (1988).

[Mi1] V. D. Milman. Asymptotic properties of functions of several variables that are defined on homogeneous spaces. Dokl. Akad. Nauk SSSR 199, 1247–1250 (1971).

[Mi2] V. D. Milman. A certain property of functions defined on infinite dimensional manifolds. Dokl. Akad. Nauk SSSR 200, 781–784 (1971).

[Mi3] V. D. Milman. New proof of the theorem of Dvoretzky on sections of convex bodies. Funct. Anal. Appl. 5, 28–37 (1971).

[Mi4] V. D. Milman. Diameter of a minimal invariant subset of equivariant Lipschitz actions on compact subsets of $\mathbb{R}^k$. Geometric Aspects of Functional Analysis, Israel Seminar 1996–2000. Lecture Notes in Math. 1267, 13–20 (1987). Springer.

[Mi5] V. D. Milman. The heritage of P. Lévy in geometrical functional analysis. Colloque Paul Lévy sur les processus stochastiques. Astérisque 157-158, 273–302 (1988).

[Mi6] V. D. Milman. Dvoretzky's theorem—Thirty years later (Survey). Geom. funct. anal. 2, 455–479 (1992).

[Mi7] V. D. Milman. Topics in asymptotic geometric analysis. Proceedings of "Visions in Mathematics—Towards 2000". GAFA 2000. Special Volume, 792–815 (2000).

[M-S] V. D. Milman, G. Schechtman. Asymptotic theory of finite dimensional normed spaces. Lecture Notes in Math. 1200 (1986). Springer.

[M-R] M. Molloy, B. A. Reed. Graph colouring by the probabilistic method (2000). In preparation.

[Nu] D. Nualart. The Malliavin calculus and related topics. Springer (1995).

[O-V] F. Otto, C. Villani. Generalization of an inequality by Talagrand, and links with the logarithmic Sobolev inequality. J. Funct. Anal. 173, 361–400 (2000).

[Os] R. Osserman. The isoperimetric inequality. Bull. Amer. Math. Soc. 84, 1182–1238 (1978).

[Pa] V. Paulauskas. Some remarks on deviation inequalities for infinitely divisible random vectors (2000).

[Pe1] V. Pestov. Amenable representations and dynamics of the unit sphere in an infinite dimensional Hilbert space. Geom. funct. anal. 10, 1171–1201 (2000).

[Pe2] V. Pestov. Ramsey-Milman phenomenon, Urysohn metric spaces and extremely amenable groups (2000). Israel J. Math., to appear.

[Pin] M. S. Pinsker. Information and information stability of random variables and processes. Holden-Day, San Franscico (1964).

[Pis1] G. Pisier. On the dimension of the ℓ_p^n-subspaces of Banach spaces, for $1 \leq p < 2$. Trans. Amer. Math. Soc. 276, 201–211 (1983).

[Pis2] G. PISIER. Probabilistic methods in the geometry of Banach spaces. Probability and Analysis, Varenna (Italy) 1985. Lecture Notes in Math. 1206, 167–241 (1986). Springer.

[Pis3] G. PISIER. The volume of convex bodies and Banach space geometry. Cambridge Univ. Press (1989).

[Ra1] S. T. RACHEV. The Monge-Kantorovich mass transference problem and its stochastic applications. Theory Probab. Appl. 24, 647–671 (1984).

[Ra2] S. T. RACHEV. Probability metrics and the stability of stochastic models. Wiley (1991).

[R-R] S. T. RACHEV, L. RÜSCHENDORFF. Mass transportation problems, vol. I & II. Springer (1998).

[R-Y] D. REVUZ, M. YOR. Continuous martingales and Brownian motions. Springer (1991).

[Ri1] E. RIO. Inégalités de Hoeffding pour les fonctions lipschitziennes de suites dépendantes. C. R. Acad. Sci. Paris 330, 905–908 (2000).

[Ri2] E. RIO. Inégalités de concentration pour les processus empiriques de classes de parties. Probab. Theory Relat. Fields 119, 163–175 (2001).

[Ri3] E. RIO. Une inégalité de Bernstein pour les maxima de processus empiriques (2001). Ann. Inst. H. Poincaré, to appear.

[Rot1] O. ROTHAUS. Diffusion on compact Riemannian manifolds and logarithmic Sobolev inequalities. J. Funct. Anal. 42, 358–367 (1981).

[Rot2] O. ROTHAUS. Logarithmic Sobolev inequalities and the growth of L^p norms. Proc. Amer. Math. Soc. 126, 2309–2314 (1998).

[Roy] G. ROYER. Une initiation aux inégalités de Sobolev logarithmiques. Cours Spécialisés. Soc. Math. de France (1999).

[Ru] W. RUDIN. Real and complex analysis. McGraw-Hill (1987).

[SC] L. SALOFF-COSTE. Lectures on finite Markov chains. Ecole d'Eté de Probabilités de St.-Flour 1996. Lecture Notes in Math. 1665, 301–413 (1997). Springer.

[Sa] P.-M. SAMSON. Concentration of measure inequalities for Markov chains and Φ-mixing processes. Ann. Probab. 28, 416–461 (2000).

[Sche1] G. SCHECHTMAN. Lévy type inequality for a class of metric spaces. Martingale Theory and Harmonic Analysis and Banach spaces, Cleveland 1981. Lecture Notes in Math. 939, 211–215 (1981). Springer.

[Sche2] G. SCHECHTMAN. A remark concerning the dependence on ϵ in Dvoretzky's theorem. Geometric Aspects of Functional Analysis (Israel Seminar, 1987–88). Lecture Notes in Math. 1376, 274–277 (1989). Springer.

[Sche3] G. SCHECHTMAN. More on embedding of Euclidean subspaces of L_p in ℓ_r^n. Compositio Math. 61, 159–170 (1987).

[Sche4] G. SCHECHTMAN. An editorial comment on the preceding paper. Geometric Aspects of Functional Analysis, Israel Seminar 1996–2000. Lecture Notes in Math. 1745, 19–20 (2000). Springer.

[Sche5] G. SCHECHTMAN. Concentration, results and applications. The Handbook in the Geometry of Banach Spaces. Elsevier (2001).

[S-S1] G. SCHECHTMAN, M. SCHMUCKENSCHLÄGER. Another remark on the volume of the intersection of two L_p^n balls. Geometric Aspects of Functional Analysis (Israel Seminar, 1989–90). Lecture Notes in Math. 1469, 174–178 (1991). Springer.

[S-S2] G. SCHECHTMAN, M. SCHMUCKENSCHLÄGER. A concentration inequality for harmonic measures on the sphere. Geometric Aspects of Functional Analysis (Israel Seminar, 1992–1994). Oper. Theory Adv. Appl. 77, 255–273 (1995). Birkhäuser.

[S-Z1] G. SCHECHTMAN, J. ZINN. On the volume of the intersection of two L_p^n. Proc. Amer. Math. Soc. 110, 217–224 (1990).

[S-Z2] G. SCHECHTMAN, J. ZINN. Concentration on the ℓ_p^n ball. Geometric Aspects of Functional Analysis, Israel Seminar 1996–2000. Lecture Notes in Math. 1745, 245–256 (2000). Springer.

[Schmi] E. SCHMIDT. Die Brunn-Minkowskische Ungleichung und ihr Spiegelbild sowie die isoperimetrische Eigenschaft der Kugel in der euklidischen und nichteuklidischen Geometrie. Math. Nach. 1, 81–157 (1948).

[Schmu1] M. SCHMUCKENSCHLÄGER. A concentration of measure phenomenon on uniformly convex bodies. Geometric Aspects of Functional Analysis (Israel Seminar, 1992–1994). Oper. Theory Adv. Appl. 77, 275–287. Birkhäuser (1995).

[Schmu2] M. SCHMUCKENSCHLÄGER. Martingales, Poincaré type inequalities and deviations inequalities. J. Funct. Anal. 155, 303–323 (1998).

[Schmu3] M. SCHMUCKENSCHLÄGER. Curvature of nonlocal Markov generators. Convex geometric analysis (Berkeley, CA, 1996), Math. Sci. Res. Inst. Publ. 34, 189–197. Cambridge Univ. Press (1999).

[Si] B. SIMON. Trace ideals and their applications. Cambridge Univ. Press (1979).

[So] A. SOSHNIKOV. Universality at the edge of the spectrum in Wigner random matrices. Comm. Math. Phys. 207, 697–733 (1999).

[Ste] J. M. STEELE. Probability theory and combinatorial optimization. CBMS-NSF Regional Conference Series in Applied Mathematics 69 (1996). SIAM.

[Sto] W. F. STOUT. Almost sure convergence. Academic Press (1974).

[Str1] D. STROOCK. Logarithmic Sobolev inequalities for Gibbs states. Dirichlet forms, Varenna 1992. Lecture Notes in Math. 1563, 194–228 (1993).

[Str2] D. STROOCK. Probability theory. An analytic view. Cambridge Univ. Press (1993).

[S-V] D. STROOCK, S. VARADHAN. Multidimensional diffusion processes. Springer (1979).

[S-T] V. N. SUDAKOV, B. S. TSIREL'SON. Extremal properties of half-spaces for spherically invariant measures. J. Soviet. Math. 9, 9–18 (1978); translated from Zap. Nauch. Sem. L.O.M.I. 41, 14–24 (1974).

[Tak] M. TAKEDA. On a martingale method for symmetric diffusion processes and its applications. Osaka J. Math. 26, 605–623 (1989).

[Tal1] M. TALAGRAND. An isoperimetric theorem on the cube and the Khintchine-Kahane inequalities. Proc. Amer. Math. Soc. 104, 905–909 (1988).

[Tal2] M. TALAGRAND. Isoperimetry and integrability of the sum of independent Banach space valued random variables. Ann. Probability 17, 1546–1570 (1989).

[Tal3] M. TALAGRAND. A new isoperimetric inequality for product measure, and the concentration of measure phenomenon. Geometric Aspects of Functional Analysis (Israel Seminar, 1989–90). Lecture Notes in Math. 1469, 91–124 (1991). Springer.

[Tal4] M. TALAGRAND. A new isoperimetric inequality for product measures and the tails of sums of independent random variables. Geom. funct. anal. 1, 211–223 (1991).

[Tal5] M. TALAGRAND. Sharper bounds for Gaussian and empirical processes. Ann. Probab. 22, 28–76 (1994).

[Tal6] M. TALAGRAND. The supremum of some canonical processes. Amer. J. Math. 116, 283–325 (1994).

[Tal7] M. TALAGRAND. Concentration of measure and isoperimetric inequalities in product spaces. Publications Mathématiques de l'I.H.E.S. 81, 73–205 (1995).

[Tal8] M. TALAGRAND. A new look at independence. Ann. Probab. 24, 1–34 (1996).

[Tal9] M. TALAGRAND. New concentration inequalities in product spaces. Invent. math. 126, 505–563 (1996).

[Tal10] M. TALAGRAND. Transportation cost for Gaussian and other product measures. Geom. funct. anal. 6, 587–600 (1996).

[Tal11] M. TALAGRAND. Majorizing measures: The generic chaining. Ann. Probab. 24, 1049–1103 (1996).

[Tal12] M. TALAGRAND. The Sherrington-Kirkpatrick model: A challenge for mathematicians. Probab. Theory Relat. Fields 110, 109–176 (1998).

[Tal13] M. TALAGRAND. Mean field models for spin glasses: A first course. Ecole d'Eté de Probabilités de St.-Flour 2000. Lecture Notes in Math., to appear. Springer.

[V-W] A. N. VAN DER VAART, J. A. WELLNER. Weak convergence of empirical processes with applications to statistics. Springer (1996).

[V-D-N] D. VOICULESCU, K. DYKEMA, A. NICA. Free random variables. CRM Monograph Series. Amer. Math. Soc. (1992).

[W-W] D. L. WANG, P. WANG. Extremal configurations on a discrete torus and a generalization of the generalized Macaulay theorem. Siam J. Appl. Math. 33, 55–59 (1977).

[Wu] L. WU. A new modified logarithmic Sobolev inequality for Poisson point processes and several applications. Probab. Theory Relat. Fields 118, 427–438 (2000).

[Yo] K. YOSHIDA. Functional analysis. Springer (1980).

[Yos] N. Yoshida. The log-Sobolev inequality for weakly coupled lattice fields. Probab. Theor. Relat. Field 115, 1–40 (1999).

[Yu1] V. V. Yurinskii. Exponential bounds for large deviations. Theor. Probability Appl. 19, 154–155 (1974).

[Yu2] V. V. Yurinskii. Exponential inequalities for sums of random vectors. J. Multivariate Anal. 6, 473–499 (1976).

INDEX